Wireless Networks

Series Editor

Xuemin Shen
University of Waterloo, Waterloo, ON, Canada

The purpose of Springer's Wireless Networks book series is to establish the state of the art and set the course for future research and development in wireless communication networks. The scope of this series includes not only all aspects of wireless networks (including cellular networks, WiFi, sensor networks, and vehicular networks), but related areas such as cloud computing and big data. The series serves as a central source of references for wireless networks research and development. It aims to publish thorough and cohesive overviews on specific topics in wireless networks, as well as works that are larger in scope than survey articles and that contain more detailed background information. The series also provides coverage of advanced and timely topics worthy of monographs, contributed volumes, textbooks and handbooks.

** Indexing: Wireless Networks is indexed in EBSCO databases and DPLB **

More information about this series at http://www.springer.com/series/14180

Peng Yang • Wen Wu • Ning Zhang • Xuemin Shen

Millimeter-Wave Networks

Beamforming Design and Performance Analysis

 Springer

Peng Yang
Huazhong University of Science and
Technology
School of Electronic Information and
Communications
Wuhan, Hubei, China

Wen Wu
Department of Electrical and Computer
Engineering
University of Waterloo
Waterloo, ON, Canada

Ning Zhang
Department of Electrical and Computer
Engineering
University of Windsor
Windsor, ON, Canada

Xuemin Shen
Department of Electrical and Computer
Engineering
University of Waterloo
Waterloo, ON, Canada

ISSN 2366-1186   ISSN 2366-1445   (electronic)
Wireless Networks
ISBN 978-3-030-88632-5   ISBN 978-3-030-88630-1   (eBook)
https://doi.org/10.1007/978-3-030-88630-1

This Springer imprint is published by the registered company Springer Nature Switzerland AG
The registered company address is: Gewerbestrasse 11, 6330 Cham, Switzerland

Preface

Millimeter-wave (mmWave) communication at 30 GHz–300 GHz frequency bands has emerged as one of the most promising technologies in future wireless networks, which can offer high data rate connections by exploiting a large swath of spectrum. The mmWave communication can support many data-intensive wireless applications, ranging from high-definition mobile video streaming, cordless virtual reality gaming, to wireless fiber-to-home access. In particular, the current IEEE 802.11ad standard can provide a data rate up to 6.75 Gbps, and its successor IEEE 802.11ay can even support up to 40 Gbps. In mmWave communications, beamforming focusing the radio frequency power in a narrow direction is the key technology to overcome the hostile path loss. However, the distinct high directionality feature of beamforming technology poses many challenges in different network layers: (1) *beam alignment (BA) latency* in the physical layer, which is the processing delay that both the transmitter and receiver take to align their beams to establish a reliable connection. Existing BA methods incur significant latency on the order of seconds for a large number of beams; (2) *medium access control (MAC) performance degradation*. To coordinate the beamforming training (BFT) for multiple users, 802.11ad standard specifies an MAC protocol, i.e., BFT-MAC protocol, in which all the users contend for BFT resources in a distributed manner. Due to the "deafness" problem caused by directional transmission, i.e., a user may not sense the transmission of other users, severe collisions occur in high user density scenarios, which significantly degrades the MAC performance; and (3) *backhaul congestion* in the network layer. All the base stations (BSs) in mmWave networks are connected to the backbone network via backhaul links to access remote content servers. Although beamforming can increase the data rate of the fronthaul links between users and BSs, the congested backhaul link becomes a new bottleneck for mmWave networks.

In this monograph, we design novel beamforming technologies for low-latency and cost-effective mmWave networks and analyze their performance. Specifically, we focus on addressing the above challenges, respectively, by (1) presenting an efficient BA algorithm, (2) evaluating and enhancing the 802.11ad MAC performance, and (3) designing an effective backhaul alleviation scheme. In Chap. 1, we introduce mmWave communications, including its definition, potential applications, recent development, and technical challenges, etc. In Chap. 2, we review the characteristics of mmWave communications, state-of-the-art beamforming technologies, and beamforming training protocol in 802.11ad. In Chap. 3, to reduce BA latency, we present a learning-based efficient BA algorithm, namely HBA. The presented algorithm leverages the correlation structure among beams and the prior knowledge on the channel fluctuation to significantly reduce BA latency from hundreds of milliseconds to a few milliseconds. In Chap. 4, to enhance the MAC performance, we present an analytical model for the BFT-MAC protocol in 802.11ad and then introduce an enhancement scheme to optimize its performance. In addition, to support multiuser transmission, we design a novel 802.11ad-compliant multiuser beamforming training protocol to reduce beamforming training overhead. In Chap. 5, to alleviate the backhaul congestion, we present a device-to-device assisted cooperative edge caching scheme that jointly utilizes cache resources of users and BSs. We theoretically analyze the performance of the introduced scheme, taking the network density, practical directional antenna model, and stochastic information of network topology into consideration. The presented scheme can effectively reduce backhaul traffic and content retrieval delay. At last, we conclude this monograph and discuss some important future research directions in Chap. 6.

We hope that this monograph can provide insightful lights on understanding the fundamental performance of mmWave networks from the perspectives of different network layers, including BA, MAC, and backhaul. The systematic principle in this monograph also offers valuable guidance on the establishment and optimization of future mmWave networks. We would like to thank Prof. Nan Cheng at the Xidian University, Prof. Khalid Aldubaikhy at the Qassim University, Prof. Yujie Tang at the Algoma University, and Prof. Weihua Zhuang at the University of Waterloo for their contributions in this monograph. We also would like to thank all the members of BBCR group for the valuable discussions and their insightful suggestions, ideas, and comments. Special thanks are also due to the staffs at Springer Science+Business Media, Susan Lagerstrom-Fife, and Arun Siva Shanmugam for their help throughout the publication preparation process.

Wuhan, Hubei, China   Peng Yang

Waterloo, ON, Canada   Wen Wu

Windsor, ON, Canada   Ning Zhang

Waterloo, ON, Canada   Xuemin Shen

Contents

1 Introduction 1
  1.1 mmWave Communication 1
      1.1.1 Motivation of mmWave Communication 1
      1.1.2 Concept and Applications 2
      1.1.3 Development of mmWave Communication 3
  1.2 Industry Progress, Projects, and Standardization 4
      1.2.1 Industry Progress 4
      1.2.2 Research Projects 5
      1.2.3 Standardization 5
  1.3 Research Challenges in mmWave Networks........................... 7
      1.3.1 Physical Layer 8
      1.3.2 MAC Layer 8
      1.3.3 Network Layer 9
  1.4 Aim of the Monograph... 10
  References 12

2 Literature Review of mmWave Networks 15
  2.1 Characteristics of mmWave Communication 15
      2.1.1 Large Bandwidth... 15
      2.1.2 Huge Path Loss 16
      2.1.3 Sparse Channel... 19
      2.1.4 Directional Antennas 20
      2.1.5 Blockage Effect 20
  2.2 Beamforming Technology 21
      2.2.1 Digital Beamforming 22
      2.2.2 Analog Beamforming.. 22
      2.2.3 Hybrid Beamforming 23
  2.3 Beamforming Training Protocol in IEEE 802.11ad 24
      2.3.1 Overview of 802.11ad Beamforming Training Protocol 25
      2.3.2 Sector Level Sweep.. 26
      2.3.3 Beam Refinement Protocol.................................... 28

	2.3.4	Beam Searching Complexity	30
2.4	Multi-armed Bandit Theory		31
2.5	Summary		33
References			34

3 Machine Learning-Based Beam Alignment in mmWave Networks 37
	3.1	Introduction	37
	3.2	Related Works on Beam Alignment	39
	3.3	System Model and Problem Formulation	41
		3.3.1 Beam Alignment Model	41
		3.3.2 Problem Statement	42
	3.4	Fast Beam Alignment Scheme	44
		3.4.1 Correlation Structure Among Beams	44
		3.4.2 Prior Knowledge of mmWave Networks	46
		3.4.3 Learning-Based Beam Alignment Algorithm	47
	3.5	Theoretical Analysis	51
		3.5.1 Algorithm Complexity Analysis	51
		3.5.2 Cumulative Regret Performance Analysis	52
	3.6	Performance Evaluation	60
		3.6.1 Simulation Setup	60
		3.6.2 Cumulative Regret	62
		3.6.3 Measurement Complexity and Beam Detection Accuracy ...	64
		3.6.4 Beam Alignment Latency	69
	3.7	Summary	70
	References		70

4 Beamforming Training Protocol Design and Analysis 73
	4.1	Introduction	73
	4.2	Existing Works on Beamforming Training	76
		4.2.1 Beamforming Training Schemes	76
		4.2.2 MAC Performance Analysis	77
	4.3	Beamforming Training Protocol in 802.11ad	78
		4.3.1 Beamforming Training Procedure	78
		4.3.2 BFT-MAC Protocol	80
	4.4	Performance Analysis and Enhancement for BFT-MAC	82
		4.4.1 Analytical Model for BFT-MAC	82
		4.4.2 Performance Analysis	85
	4.5	Performance Evaluation for 802.11ad BFT-MAC	95
		4.5.1 Simulation Setup	95
		4.5.2 Validation of Analytical Model	96
		4.5.3 Enhancement Scheme Evaluation	99
	4.6	Multiuser Beamforming Training Protocol Design and Analysis	103
		4.6.1 Existing Works on Multiuser Transmission	104
		4.6.2 Multiuser Transmission Scheme	104
		4.6.3 Multiuser Beamforming Training Protocol	108
		4.6.4 Protocol Overhead Analysis	110

4.7   Protocol Performance Evaluation 112
     4.7.1   Simulation Setup.. 113
     4.7.2   Simulation Results 114
4.8   Summary 117
References 117

5   Beamforming-Aided Cooperative Edge Caching in mmWave
Dense Networks 121
5.1   Introduction 121
5.2   Related Works on Edge Caching 122
5.3   System Model 124
     5.3.1   Network Model 125
     5.3.2   Content Popularity Model 126
     5.3.3   Directional Antenna Model 126
     5.3.4   mmWave Channel Model...................................... 127
     5.3.5   Transmission Model 127
5.4   D2D-Assisted Cooperative Edge Caching (DCEC) Policy 128
     5.4.1   Scheme Design... 129
     5.4.2   Backhaul Offloading Analysis 130
5.5   Content Retrieval Delay Analysis 131
     5.5.1   Backhaul Transmission Rate Analysis......................... 131
     5.5.2   Nearest SBS Transmission Rate Analysis 132
     5.5.3   SBS Cluster Transmission Rate Analysis...................... 137
     5.5.4   D2D Transmission Rate Analysis.............................. 140
5.6   Performance Evaluation 143
     5.6.1   Simulation Setup... 143
     5.6.2   Backhaul Offloading Performance 144
     5.6.3   Transmission Performance 146
     5.6.4   Content Retrieval Delay 148
5.7   Summary 153
References 153

6   Summary and Future Directions 157
6.1   Summary 157
     6.1.1   Beam Alignment Scheme Design 157
     6.1.2   MAC Performance Evaluation and Enhancement............. 158
     6.1.3   Backhaul Alleviation Scheme Design 158
6.2   Future Directions.. 159
     6.2.1   Beam Alignment Under High Mobility........................ 159
     6.2.2   Efficient QoS-Aware MAC Protocol........................... 159
     6.2.3   Blockage-Aware mmWave Network 160

Acronyms

3GPP	3rd generation partnership project
5G	Fifth generation
6G	Sixth generation
A-BFT	Associated beamforming training
ADC	Analog-to-digital conversion
AOA	Angle of arrival
AOD	Angle of departure
AP	Access point
AR	Augmented reality
ATI	Announcement transmission interval
BA	Beam alignment
BC	Beam combining
BF	Beamforming
BFT	Beamforming training
BFT-MAC	Beamforming training medium access control
BHI	Beacon header interval
BI	Beacon interval
BRP	Beam refinement protocol
BS	Base station
BT	Beam tracking
BTI	Beacon transmission interval
CBAP	Contention-based access period
CMOS	Complementary metal-oxide semiconductor
COTS	Commercial-off-the-shelf
D2D	Device-to-device
DAC	Digital-to-analog conversion
DCEC	Device-to-device assisted cooperative edge caching
DCF	Distributed coordinate function
DFT	Discrete Fourier transform
DSP	Digital signal processing
DTI	Data transmission interval

EIRP	Effective isotropically radiated power
eMBB	Enhanced mobile broadband
FCC	Federal communications commission
FWA	Fixed wireless access
HBA	Hierarchical beam alignment
HOO	Hierarchical optimistic optimization
ID	Identification
IoT	Internet of things
ISS	Initial sector sweep
LOS	Line-of-sight
MAB	Multi-armed bandit
MAC	Medium access control
MEC	Mobile edge computing
MID	Multiple sector ID
MIMO	Multiple input multiple output
mmWave	Millimeter wave
mMTC	Massive machine type communications
MPC	Most popular caching
NLOS	Non-line-of-sight
PDF	Probability distribution function
PPP	Poisson point process
QoE	Quality of experience
QoS	Quality of service
RF	Radio frequency
RSS	Received signal strength
SBS	Small base station
SINR	Signal-to-interference-plus-noise ratio
SLS	Sector level sweep
SNR	Signal-to-noise ratio
SP	Service period
SPR	Service period request
SSW	Sector sweep
SSW-FB	Sector sweep feedback
STA	Station
TCP	Transmission control protocol
TDMA	Time division multiple access
UBA	Unimodal beam alignment
UCB	Upper confidence bound
UDN	Ultra-dense network
URLLC	Ultra-reliable low-latency communication
VR	Virtual reality
WLAN	Wireless local area network
WPAN	Wireless personal area network

Chapter 1
Introduction

1.1 mmWave Communication

1.1.1 Motivation of mmWave Communication

Recently, we have witnessed the breathtaking progress of advanced technologies in consumer electronic devices and the surge of a number of data-intensive applications, ranging from panoramic video streaming, big data analytics, wireless fiber-to-home access, cordless virtual reality (VR), to mobile augmented reality (AR) gaming [1]. These emerging applications provide people with ubiquitous high-quality multimedia content. It is reported that HTC has sold more than 15,000 VR headsets in just 10 min on its first release. Another report shows that the worldwide shipment of AR/VR devices reached 9 million in 2019 [2]. These advanced devices and the corresponding data-intensive applications are expected to push a significant growth of data volume in wireless networks. As predicted, the monthly global mobile data traffic is expected to have a 12-fold increase in the next 5 years [3].

Traditional wireless networks that operate in congested microwave frequency bands (i.e., below 6 GHz) face a severe spectrum scarcity issue, especially at peak hours. It is difficult to satisfy the surging demand of mobile data traffic purely relying on traditional wireless networks. To address this issue, i.e., supporting these data-intensive applications, it is necessary that wireless networks should deliver a multiple gigabit data rate [4]. Such a requirement is unquestionably beyond the capability of traditional wireless networks, even if high spectrum efficiency techniques are utilized regardless of their complexity. Therefore, to meet the requirement, more spectrum is needed in future wireless networks.

© The Author(s), under exclusive license to Springer Nature Switzerland AG 2021
P. Yang et al., *Millimeter-Wave Networks*, Wireless Networks,
https://doi.org/10.1007/978-3-030-88630-1_1

1.1.2   Concept and Applications

High-frequency band communication has garnered great attention from both academia and industry due to the spectrum scarcity issue in wireless communication. In particular, communication ranging from 30 GHz to 300 GHz frequency bands, whose wavelength ranges from 1 mm to 10 mm, becomes the focus. This range of frequency bands is often referred to as millimeter-wave (mmWave) communication, which is in possession of a large swath of spectrum. Based on federal communications commission (FCC) regulations, more than 20 GHz of spectrum are available at the mmWave frequency band. The amount of spectrum is 20 times more than that allocated to both current Wi-Fi and cellular systems. Leveraging such abundant spectrum, mmWave communication can provide a multiple gigabit data rate. Hence, mmWave communication is one of the key enabling technologies in the advanced wireless networks, such as the fifth-generation (5G) networks [5–8], and is expected to continue playing an important role in the coming 6G networks [9–12].

Due to the benefit of high data rate, mmWave communications can support numerous potential data-hungry applications. (1) Supporting infotainment applications is an important use case, in which mmWave communication can enhance the quality of experience (QoE) of mobile users. For instance, mobile users can enjoy high-quality VR/AR gaming or watch real-time ultra high-definition videos through mmWave communication. (2) From the perspective of network operators, mmWave communication can be an enabling technology for wireless backhaul in ultradense networks (UDNs) and emerging drone networks. In UDNs, backhaul links need to offer a data rate up to 10 Gbps to provide reliable connections [13]. The deployment cost of high-speed wired backhaul links is high. As such, mmWave communication is a cost-effective solution for the UDNs. Similar to that in the UDNs, mmWave communication is also considered as the underlying wireless backhaul technology in drone networks. (3) In addition to the above wireless backhaul applications, mmWave communication can also be applied to solve the issue of last mile optical fiber replacement. Recently, the emerging fixed wireless access technology which utilizes mmWave communication to provide broadband Internet access to customers seems plausible. This scheme can avoid the costly and time-consuming deployment of optical fiber networks [14], and (4) mmWave communication can be applied in high-mobility scenarios. For example, it can be used as the access network of high-speed railways to support a large number of users in the train [15]. Also, mmWave communication can be employed in vehicular networks, which can not only offer high-speed vehicle to everything communications [16, 17] but also sense moving objects in nearby environment for safety driving owing to its short signal wavelength [18]. To summarize, mmWave communication can facilitate a large number of high-speed data services for wireless networks in a cost-effective manner.

1.1.3   Development of mmWave Communication

The earliest concept of mmWave communication can be traced back to more than 100 years ago. In 1897, Bose conducted the first mmWave experiment with a wavelength of 5 mm [19], while in the following century, mmWave communication was deemed unsuitable for wireless communications. The reason is threefold: (1) As compared to signal propagation at the microwave band, signal propagation at mmWave frequency bands suffers from significant path loss; (2) due to weak reflection characteristics, the mmWave channel is relatively sparse. The insufficiency of available paths limits the coverage of mmWave networks; and (3) mmWave signals with a short signal wavelength suffers from huge penetration loss, and hence they are extremely vulnerable to blockage. The above hostile transmission characteristics significantly hinder the development of mmWave communication in the past century.

With the recent progress of advanced hardware circuit, wireless communication, and signal processing technology, mmWave communication is no longer conceptual nowadays. Utilizing advanced low-power complementary metal-oxide-semiconductor (CMOS) radio-frequency (RF) circuits technologies and the small wavelength of mmWave signals, a large number of antennas (e.g., 32 antenna elements) are packed into an antenna array of a compact size, thereby addressing the huge path loss issue. These compact yet energy-efficient antenna arrays can be fabricated in a small chip, which facilitate lightweight and long-life mmWave communication at mobile devices. The antenna array can provide high directional antenna gain, namely, *beamforming*. In specific, beamforming acts as a "focusing lens." It can focus RF energy toward a narrow direction to enhance the received signal-to-noise ratio (SNR) at mmWave frequency bands. For instance, theoretical analysis shows that an antenna array with 32 antenna elements can provide around 15 dB antenna gain. The antenna array at large-space and energy-abundant BSs can even adopt a larger number of antennas, such as 256 or 1024 antenna elements, to provide a higher antenna gain. Thus, with the beamforming technology, the issue of the path loss can be addressed in mmWave communication.

To promote mmWave communication, extensive research works in academia are conducted, including studying mmWave network performance [20], designing novel efficient beamforming technologies [21, 22], and enhancing network performance from perspectives of different network layers, ranging from the physical layer, the medium access control (MAC) layer, to the network layer. In addition, there are also extensive efforts from the industry, which are introduced in details in the following section.

1.2 Industry Progress, Projects, and Standardization

Extensive efforts from industry are devoted to large-scale in-field tests and standardization to pave the way for commercialization of mmWave communications. Recent progress, research projects, and existing standards are reviewed as follows.

1.2.1 Industry Progress

Since mmWave communication is still at the infancy stage, network operators and vendors perform extensive large-scale field trials to evaluate the performance of mmWave communication at different frequency bands under various settings. It is reported that Samsung conducted the first mmWave mobile communication experiment in 2013.[1] The achieved data rate is up to 1 Gbps. Other network operators, i.e., T-Mobile and Verizon, test mmWave communication at 28 and 39 GHz bands with the permission of FCC. In another experiment joint performed by Nokia and National Instrument, the achieved data rate is up to 15 Gbps at the 73 GHz band. Huawei also performs an experiment at mmWave frequency bands, i.e., the Ka band from 26.5 GHz to 40 GHz. The experimental results show a 20 Gbps access rate for mobile users. Another experiment by DOCOMO and Ericsson at the 70 GHz band shows that the data rate can reach 4.5 Gbps and 2 Gbps in outdoor and indoor scenarios, respectively. The above field trials validate that mmWave communication is able to offer high-speed data transmission.

In addition, numerous measurements on channel conditions and network coverage have been conducted by standard bodies, such as the 3rd Generation Partnership Project (3GPP) and Wi-Fi Alliance. Several 3GPP working groups have constructed empirical channel models for mmWave communication based on extensive measurement data in different scenarios. Some research groups also conduct important measurements in some use cases, such as urban communication scenarios. We know that mmWave communication would suffer from significant penetration loss in dense urban scenarios. This is because the dense buildings may result in a coverage concern of mmWave networks. To validate the feasibility of mmWave networks, extensive experiments are conducted by the NYU WIRELESS research group in New York City at both 28 GHz and 38 GHz bands. The measurement takes signal penetration and reflection characteristics at the buildings into consideration. The measurement results show that, even in a dense urban environment with a low-power base station (BS), the mmWave networks still have a coverage up to 200 m [23].

---

[1] Online: https://news.samsung.com/global/samsung-announces-worlds-first-5g-mmwave-mobile-technology.

1.2.2   Research Projects

To further push the development of mmWave mobile networks, industry and government institutions all over the world have launched multiple research projects on mmWave networks. For example, a joint project by European union (EU) and Japan, named "MiWEBA",[2] aims at the mmWave evolution for backhaul and access networks. Other projects, such as "MiWaves"[3] and "mmMAGIC",[4] are also launched by EU to push the development of mmWave communication. Not surprisingly, other countries, such as China and America, have invested hundreds of millions of dollars on mmWave communication research and implementation [24]. Chinese governments have launched a few key research projects on studying mobile mmWave networks and designing corresponding RF chips. Leading technology companies in the industry, such as Google, Facebook, and Huawei, also put extensive efforts into this topic. For example, Google has tested 71–76 GHz and 81–86 GHz frequency bands for potential wireless backhaul applications.

1.2.3   Standardization

Standardization efforts are also made for the commercialization of mmWave communication with the focus on both unlicensed and licensed bands, and a number of standard-compliant products hit the market in recent years.

1.2.3.1   Standards at the Unlicensed Bands

In recent years, multiple standards have been ratified for diverse applications by different standard groups at the unlicensed 57–64 GHz bands, often referred to as the unlicensed 60 GHz band in the literature. These standards include WirelessHD for video area networks,[5] IEEE 802.15.3c for wireless personal area networks (WPANs) [25], IEEE 802.11ad (refers to 802.11ad for short hereinafter) for wireless local area networks (WLANs) [26], and 802.11ad's successor IEEE 802.11ay [27]. We introduce these standards as follows:

- *WirelessHD*: Silicon image consortium develops WirelessHD to provide high-definition video transmission via short-range mmWave communication. The achievable data rate is up to 4 Gbps. Such a data rate can support common 3D formats and 4K resolution video transmission.

---

[2] MiWEBA: http://www.miweba.eu/.

[3] MiWaves: http://www.miwaves.eu/.

[4] mmMAGIC: https://5g-mmmagic.eu/.

[5] WirelessHD: https://en.wikipedia.org/wiki/WirelessHD.

- *IEEE 802.15.3c*: The standard is designed to provide mmWave communication for the existing 802.15.3 WPANs. The standard can deliver a data rate up to 2 Gbps for supporting data-hungry applications, such as high-definition TV video streaming and wireless data bus for cable replacement.
- *IEEE 802.11ad*: The standard was developed in the former WiGig consortium that was later absorbed into the Wi-Fi Alliance in 2012. In 802.11ad, nearly 7 GHz spectrum at the unlicensed 60 GHz band can be leveraged to provide a data rate up to 6.75 Gbps with a channel bandwidth of 2.16 GHz. FCC released extra 7 GHz unlicensed spectrum at the 64–71 GHz band to further promote 802.11ad in 2016. The new released spectrum doubles the amount of available spectrum in 802.11ad. In this way, 802.11ad can occupy at most six channels, thereby supporting more users and providing a higher data rate.
- *IEEE 802.11ay*: Seeing the great success of 802.11ad, IEEE standardization group is developing the next-generation standard for mmWave WLANs, named 802.11ay. Compared with 802.11ad, numerous distinguished technologies, such as channel bonding, channel aggregation, and multiuser transmission, would be incorporated in 802.11ay [28, 29], which can significantly increase data rate up to 40 Gbps. The draft version of 802.11ay debut in 2019. For more details on the 802.11ay, the readers are referred to a comprehensive survey in [30].

1.2.3.2 Standards at the Licensed Bands

Standardization activities at the *licensed band* are led by 5G-related forums and organizations. The 5G networks support three typical use cases: (i) enhanced mobile broadband (eMBB) that requires extremely high data rates, (ii) massive machine-type communications (mMTC) that require massive connectivity in Internet of things (IoT), and (iii) ultra-reliable low-latency communication (URLLC) that focuses on applications with stringent latency and reliability requirements [5]. The peak data rate in the eMBB use case is up to 10 Gbps. To satisfy this requirement, mmWave communication is unquestionably the most promising technology. To bring mmWave visions to commercialization, the 3GPP working group has standardized mmWave communication as a 5G new radio interface in Release 15. The mmWave communication would continue evolving in the subsequent releases, including 5G beyond and the 6G networks.

1.2.3.3 Commercial Products

Many standard-compliant commercial off-the-shelf (COTS) products hit the market in recent years. WirelessHD-compliant products have been available for years. For 802.11ad, building on the successful 2.4/5 GHz Wi-Fi systems, it achieved significant success in the past years. Many COTS products are designed based on 802.11ad. Very recently, a group of companies, such as Wilocity, Tensorcom, Netgear, and Nitero [31], can offer 802.11ad-capable chips. For instance, two

Table 1.1 Comparison between 802.11ac-based Wi-Fi and 802.11ad-based mmWave communication

Metric	Wi-Fi (802.11ac)	mmWave communication (802.11ad)
Bandwidth	160 MHz	2.16 GHz
Peak data rate	500 Mbit/s	6.75 Gbps
Carrier frequency	5 GHz	60 GHz
Path loss	Medium	Severe
Number of antennas	1–8	16–32
Antenna directionality	Omni-directional	Directional

802.11ad-capable routers hit the market, i.e., Netgear Nighthawk X10 and TP-Link Talon AD7200 [32]. As reported by the ABI research [33], the market of mmWave WLAN is booming, and more than 600 million 802.11ad-compliant Wi-Fi chips has been shipped in 2020. The ongoing standardization of 802.11ay can further promote the development of mmWave communication in WLANs.

Based on existing standards, we compare the common Wi-Fi technology (state-of-the-art 802.11ac) and mmWave communication (802.11ad) to have a better understanding of mmWave communication. The detailed comparison is presented in Table 1.1. Compared with Wi-Fi, mmWave communication possesses multiple GHz bandwidth, and hence it can achieve a higher data rate than Wi-Fi. The peak data rate of mmWave communication is nearly 7 Gbps. In addition, unlike Wi-Fi which operates in the microwave band, i.e., 5 GHz band, mmWave communication operates in the 60 GHz band. Due to operating at a high-frequency band, the path loss in mmWave communication is severe. Thus, to compensate the path loss, a large number of antennas are equipped to form directional antennas in mmWave communication. The directional antenna is the main difference between Wi-Fi and mmWave communication, which significantly impacts scheme design and performance analysis in mmWave communication.

1.3   Research Challenges in mmWave Networks

Beamforming that focuses the radio-frequency power in a narrow direction is the key technology in mmWave communication to overcome the hostile path loss. The *high-directionality* feature brought by beamforming poses new research challenges on mmWave networks from perspectives of different network layers, including physical layer, MAC layer, and network layer, which are detailed as follows.

Fig. 1.1 An illustration of
beam alignment in the
physical layer, which aims to
identify the optimal beam pair
between the transmitter and
the receiver among a number
of beam combinations

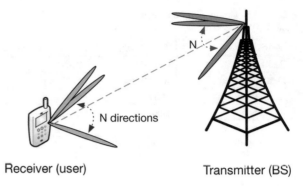

1.3.1 Physical Layer

Due to the high-directionality feature endowed by beamforming, both the transmitter and the receiver need to accurately align their beams to establish a reliable
communication link in the physical layer, i.e., *beam alignment*.[6] As an illustrative
example shown in Fig. 1.1, a beam alignment process is required to establish a
reliable communication link between users and the BS. Even with slight beam
misalignment, link budget can be dramatically reduced, and the throughput can drop
from multiple gigabits to a few hundred megabits [34]. Therefore, beam alignment
is a key process for achieving high-speed data transmission in mmWave networks.

The objective of beam alignment is to identify the optimal transmit-receive beam
pair among all possible beam combinations. However, existing methods, including
those in 802.11ad, have to search the entire beam space until the optimal transmit-
receive beam pair is identified. These methods incur significant beam alignment
latency, which can be at the order of seconds when the number of beams is large.
Hence, it is desired to have an efficient beam alignment algorithm without searching
the entire beam space which can be applied in practical mmWave networks.

1.3.2 MAC Layer

In the MAC layer, beamforming training is a challenging issue. Beamforming
training should be performed for all the users to associate to the mmWave network.
The beamforming training is operated under the coordination of a BS or an access
point (AP), as shown in Fig. 1.2. To carry out beamforming training for multiple
users in the mmWave network, 802.11ad specifies a distributed beamforming

---

[6] In our monograph, the terminology "beam alignment" is used in the physical layer, while
the terminology "beamforming training" is used in the MAC layer since 802.11ad adopts the
terminology "beamforming training."

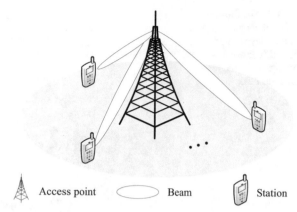

Fig. 1.2 The illustration of beamforming training in the MAC layer, in which all users contend for the limited beamforming training opportunities

training MAC protocol, namely, BFT-MAC. In the protocol, each user performs beamforming training in a contention and back-off manner. Due to the high-directionality feature in mmWave communication, a "deafness" problem is incurred, in which a user may not sense the transmission of other users. As such, severe transmission collisions would occur in high user density scenarios. Thus, it is paramount to analyze the performance of BFT-MAC, such that MAC parameters can be optimally configured to optimize MAC performance.

The BFT-MAC protocol is distinct from traditional MAC protocols with carrier sensing mechanisms (e.g., [35, 36]) due to the adoption of directional antennas. A new analytical model is desired for the BFT-MAC protocol. Moreover, it is imperative to enhance the MAC performance in high user density scenarios, in which the severe performance degradation occurs due to significant transmission collisions. In addition, since multiuser transmission would be a key feature technology in future mmWave communication, designing a beamforming training protocol for supporting multiuser transmission is desired, namely, multiuser beamforming training protocol. Also, the designed protocol should be compliant with the current 802.11ad standard for feasibility consideration.

1.3.3   Network Layer

In the network layer, a backhaul congestion issue should be addressed, especially in mmWave dense networks. As shown in Fig. 1.3, backhaul links between BSs and backbone networks are relatively congested in mmWave dense networks [37]. This is due to the prohibitive high deployment cost of ultrahigh-speed wired backhaul links. Although high data rate mmWave communication can significantly enhance the fronthaul links between users and BSs, the congested backhaul links become the bottleneck of the entire mmWave networks from the perspective of the network layer. The issue would further degrade the network performance with the

Fig. 1.3 The illustration of backhaul congestion in the network layer, in which network performance would degrade with congested backhaul links

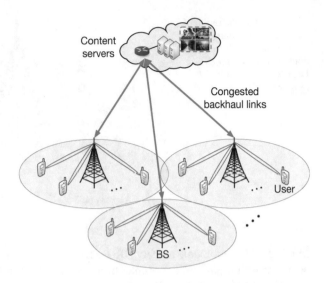

increase of the densification of mmWave networks. Therefore, it is necessary to develop an effective solution to alleviate backhaul congestion for mmWave network deployment.

1.4 Aim of the Monograph

This monograph aims to design low-latency and cost-effective schemes to enhance mmWave network performance from the perspectives of different network layers, including the physical layer, the MAC layer, and the network layer.

Firstly, in Chap. 2, we present a comprehensive survey to elaborate the unique features of mmWave networks. Then, we elaborate different types of beamforming technologies. The beamforming training protocol in the celebrated 802.11ad standard is introduced in detail. Next, some preliminaries on a specific machine learning method, i.e., multi-armed bandit (MAB), are introduced since it is adopted to address the beam alignment issue.

Secondly, in Chap. 3, to address the beam alignment challenge in the physical layer, we present a fast beam alignment algorithm, namely, hierarchical beam alignment (HBA) algorithm [38]. This algorithm can significantly reduce beam alignment latency. Specifically, the beam alignment problem is formulated as a stochastic MAB problem, with the objective of maximizing the cumulative received signal strength (RSS) within a certain time period. To solve the problem, the presented HBA algorithm takes advantage of the correlation structure among nearby beams. As such, the information from nearby beams is extracted to identify the optimal beam, thereby avoiding searching the entire beam space and reducing beam alignment latency. In addition, the presented algorithm incorporates the prior knowledge on the channel fluctuation, which can further accelerate the beam alignment process. We conduct theoretical analysis, which indicates that the presented algorithm is asymptotically optimal. Extensive simulations based on 802.11ad in the multipath channel scenarios are provided. The results demonstrate that the presented algorithm can identify the optimal beam with an extremely high probability. As compared to the existing beam alignment algorithms in 802.11ad, the presented algorithm can reduce the beam alignment latency from hundreds of milliseconds to a few milliseconds.

Thirdly, in Chap. 4, to address the beamforming training challenge in the MAC layer, we *first* analyze the BFT-MAC protocol performance in 802.11ad. Based on theoretical analysis, we then present an enhancement scheme for high user density scenarios [39, 40]. Specifically, a simple yet accurate analytical model is presented for characterizing the BFT-MAC protocol behavior. In the presented analytical model, we incorporate the user density into the modeling of the BFT-MAC protocol. The presented analytical model can unveil the impact of user density and MAC parameters on the MAC performance. Based on our analytical model, we further derive the closed-form expressions of multiple network performance metrics, including average successful beamforming training probability, the normalized throughput, and the beamforming training latency. Through asymptotic analysis, the maximum normalized throughput of BFT-MAC is proved to be barely $1/e$. The throughput is the same as that of slotted ALOHA protocol. Moreover, in dense user scenarios, due to the mismatch between the number of active users and available beamforming training resources, network throughput suffers from great degradation. Next, to solve the issue, we introduce an enhancement scheme which adaptively adjusts the MAC parameters in tune with user density. Extensive simulation are conducted, and the results show that the presented analytical model is accurate, and this enhancement scheme is effective in dense user scenarios. Second, multiuser transmission can greatly increase data rate in mmWave networks, which is expected to be a key enabling technology in the ongoing 802.11ay standard. Designing a beamforming training protocol for multiuser transmission, i.e., multiuser beamforming training protocol, is needed. We introduce a novel protocol based on the hybrid beamforming algorithm [41]. This protocol is compliant with the current 802.11ad standard, which indicates its feasibility. In addition, we analyze the protocol overhead, and simulations are conducted to validate the effectiveness of this protocol.

Fourthly, in Chap. 5, to address the backhaul congestion challenge in the network layer, we adopt edge caching as a potential solution [42–44]. Edge caching proactively caches popular contents in nearby small base stations (SBSs), which can effectively alleviate the backhaul congestion in mmWave dense networks. However, cache resources of individual SBSs are constrained, thereby significantly throttling edge caching performance. To address this issue, we present a cooperative edge caching scheme, namely, device-to-device (D2D) assisted cooperative edge caching (DCEC) [45]. The presented scheme cooperatively utilizes the cache resource of users and SBSs in proximity. Specifically, in DCEC, a popular content can be cached in either users or SBSs according to the content popularity. Then, users can request contents either from their neighboring users via high-rate D2D links or the neighboring SBSs via mmWave cellular links. As such, the cache diversity is efficiently exploited. The existing cooperative caching schemes in the microwave frequency bands require complex interference management techniques to suppress interference. While in mmWave systems, we take advantage of directional antennas to ensure high transmission rate while solving the interference issue. We derive closed-form expressions of the backhaul offloading performance and content retrieval delay based on the stochastic information of network topology, taking the practical directional antenna model and network density into account. Furthermore, analytical results show that content retrieval delay via D2D links increases significantly with the increase of network density, while that via cellular links increases slightly. Comprehensive simulation results validate the accuracy of our theoretical analysis. In addition, the results demonstrate that the presented scheme can achieve significant performance gains both in backhaul traffic offloading and content retrieval delay reduction, as compared to the most popular caching benchmark.

Finally, in Chap. 6, we summarize this monograph and discuss potential future research directions in mmWave networks, including beam alignment schemes in high-mobility scenarios, efficient QoS-aware MAC protocols, and anti-blockage methods.

References

1. W. Wu, Design and analysis of beamforming in mmwave networks, University of Waterloo, Waterloo, 2019
2. International Data Corporation, Augmented reality and virtual reality headsets poised for significant growth, 2019 [Online]. Available: https://www.idc.com/getdoc.jsp?containerId=prUS44966319
3. Cisco, Cisco visual networking index: Global mobile data traffic forecast update, 2017–2022 white paper, 2019. [Online]. Available: https://www.cisco.com/c/en/us/solutions/collateral/service-provider/visual-networking-index-vni/white-paper-c11-741490.html
4. O. Abari, D. Bharadia, A. Duffield, D. Katabi, Enabling high-quality untethered virtual reality, in *Proc. USENIX NSDI* (2017), pp. 531–544
5. J.G. Andrews, S. Buzzi, W. Choi, S.V. Hanly, A. Lozano, A.C. Soong, J.C. Zhang, What will 5G be? IEEE J. Sel. Areas Commun. **32**(6), 1065–1082 (2014)
6. N. Zhang, P. Yang, J. Ren, D. Chen, L. Yu, X. Shen, Synergy of big data and 5G wireless networks: opportunities, approaches, and challenges. IEEE Wireless Commun. **25**(1), 12–18 (2018)
7. K. Qu, W. Zhuang, Q. Ye, X.S. Shen, X. Li, J. Rao, Traffic engineering for service-oriented 5G networks with SDN-NFV integration. IEEE Netw. **34**(4), 234–241 (2020)
8. N. Zhang, N. Cheng, A.T. Gamage, K. Zhang, J.W. Mark, X. Shen, Cloud assisted HetNets toward 5G wireless networks. IEEE Commun. Mag. **53**(6), 59–65 (2015)
9. X. You et al., Towards 6G wireless communication networks: vision, enabling technologies, and new paradigm shifts. Sci. China Inf. Sci. **64**(1), 1–74 (2021)
10. X. Shen, J. Gao, W. Wu, K. Lyu, M. Li, W. Zhuang, X. Li, J. Rao, AI-assisted network-slicing based next-generation wireless networks. IEEE Open J. Veh. Technol. **1**(1), 45–66 (2020)

11. K. Letaief, W. Chen, Y. Shi, J. Zhang, Y.-J.A. Zhang, The roadmap to 6G: AI empowered wireless networks. IEEE Commun. Mag. **57**(8), 84–90 (2019)
12. W. Wu, N. Chen, C. Zhou, M. Li, X. Shen, W. Zhuang, X. Li, Dynamic RAN slicing for service-oriented vehicular networks via constrained learning. IEEE J. Sel. Areas Commun. **39**(7), 2076–2089 (2021)
13. I. Chih-Lin, C. Rowell, S. Han, Z. Xu, G. Li, Z. Pan, Toward green and soft: a 5G perspective. IEEE Commun. Mag. **52**(2), 66–73 (2014)
14. K. Aldubaikhy, W. Wu, N. Zhang, N. Cheng, X. Shen, mmwave IEEE 802.11ay for 5G fixed wireless access. IEEE Wirel. Commun. **27**(2), 88–95 (2020)
15. M. Gao, B. Ai, Y. Niu, W. Wu, P. Yang, F. Lyu, X. Shen, Efficient hybrid beamforming with anti-blockage design for high-speed railway communications. IEEE Trans. Veh. Technol. **69**(9), 9643–9655 (2020)
16. N. Lu, N. Cheng, N. Zhang, X. Shen, J.W. Mark, Connected vehicles: solutions and challenges. IEEE Internet Things J. **1**(4), 289–299 (2014)
17. N. Zhang, S. Zhang, P. Yang, O. Alhussein, W. Zhuang, X.S. Shen, Software defined space-air-ground integrated vehicular networks: challenges and solutions. IEEE Commun. Mag. **55**(7), 101–109 (2017)
18. J. Choi, V. Va, N. Gonzalez-Prelcic, R. Daniels, C.R. Bhat, R.W. Heath, Millimeter-wave vehicular communication to support massive automotive sensing. IEEE Commun. Mag. **54**(12), 160–167 (2016)
19. D.T. Emerson, The work of Jagadis Chandra Bose: 100 years of millimeter-wave research. IEEE Trans. Microw. Theory Technol. **45**(12), 2267–2273 (1997)
20. T. Bai, R.W. Heath, Coverage and rate analysis for millimeter-wave cellular networks. IEEE Trans. Wireless Commun. **14**(2), 1100–1114 (2015)
21. R.W. Heath, N. Gonzalez-Prelcic, S. Rangan, W. Roh, A.M. Sayeed, An overview of signal processing techniques for millimeter wave MIMO systems. IEEE J. Sel. Topics Signal Process. **10**(3), 436–453 (2016)
22. A. Alkhateeb, G. Leus, R.W. Heath, Limited feedback hybrid precoding for multi-user millimeter wave systems. IEEE Trans. Wireless Commun. **14**(11), 6481–6494 (2015)
23. T.S. Rappaport, S. Sun, R. Mayzus, H. Zhao, Y. Azar, K. Wang, G.N. Wong, J.K. Schulz, M. Samimi, F. Gutierrez, Millimeter wave mobile communications for 5G cellular: it will work! IEEE Access **1**, 335–349 (2013)
24. M. Xiao, S. Mumtaz, Y. Huang, L. Dai, Y. Li, M. Matthaiou, G.K. Karagiannidis, E. Björnson, K. Yang, I. Chih-Lin, Millimeter wave communications for future mobile networks. IEEE J. Sel. Areas Commun. **35**(9), 1909–1935 (2017)
25. IEEE Standards Association, IEEE standards 802.15.3c-2009: Millimeter-wave-based alternate physical layer extension, 2009
26. IEEE Standards, IEEE standards 802.11 ad-2012: enhancement for very high throughput in the 60 GHz band, 2012
27. Y. Ghasempour, C.R. da Silva, C. Cordeiro, E.W. Knightly, IEEE 802.11 ay: Next-generation 60 GHz communication for 100 Gb/s Wi-Fi. IEEE Commun. Mag. **55**(12), 186–192 (2017)
28. K. Aldubaikhy, W. Wu, Q. Ye, X. Shen, Low-complexity user selection algorithms for multiuser transmissions in mmwave WLANs. IEEE Trans. Wireless Commun. **19**(4), 2397–2410 (2020)
29. K. Aldubaikhy, W. Wu, X. Shen, HBF-PDVG: hybrid beamforming and user selection for UL MU-MIMO mmWave systems, in *Proceedings of the IEEE Globecom Workshops* (2018), pp. 1–6
30. P. Zhou, K. Cheng, X. Han, X. Fang, Y. Fang, R. He, Y. Long, Y. Liu, IEEE 802.11ay-based mmWave WLANs: design challenges and solutions. IEEE Commun. Surv. Tut. **20**(3), 1654–1681 (2018)
31. A. Loch, G. Bielsa, J. Widmer, Practical lower layer 60 GHz measurements using commercial off-the-shelf hardware, in *Proceedings of the ACM WiNTECH* (2016), pp. 9–16

32. M. Boers, B. Afshar, I. Vassiliou, S. Sarkar, S.T. Nicolson, E. Adabi, B.G. Perumana, T. Chalvatzis, S. Kavvadias, P. Sen et al., A 16TX/16RX 60 GHz 802.11 ad chipset with single coaxial interface and polarization diversity. IEEE J. Solid-State Circ. **49**(12), 3031–3045 (2014)
33. ABI Research, 802.11ad will Vastly Enhance Wi-Fi: The Importance of the 60 GHz Band to Wi-Fi's Continued Evolution, 2016 [Online]. Available: https://www.qualcomm.com/media/documents/files/abi-research-802-11ad-will-vastly-enhance-wi-fi.pdf
34. T. Wei, X. Zhang, Pose information assisted 60 Ghz networks: towards seamless coverage and mobility support, in *Proceedings of the ACM MOBICOM* (2017), pp. 42–55.
35. G. Bianchi, Performance analysis of the IEEE 802.11 distributed coordination function, IEEE J. Sel. Areas Commun. **18**(3), 535–547 (2000)
36. Q. Ye, W. Zhuang, L. Li, P. Vigneron, Traffic-load-adaptive medium access control for fully connected mobile ad hoc networks. IEEE Trans. Veh. Technol. **65**(11), 9358–9371 (2016)
37. P. Yang, N. Zhang, Y. Bi, L. Yu, X. Shen, Catalyzing cloud-fog interoperation in 5G wireless networks: an SDN approach. IEEE Netw. **31**(5), 14–20 (2017)
38. W. Wu, N. Cheng, N. Zhang, P. Yang, W. Zhuang, X. Shen, Fast mmwave beam alignment via correlated bandit learning. IEEE Trans. Wireless Commun. **18**(12), 5894–5908 (2019)
39. W. Wu, N. Cheng, N. Zhang, P. Yang, K. Aldubaikhy, X. Shen, Performance analysis and enhancement of beamforming training in 802.11ad. IEEE Trans. Veh. Technol. **69**(5), 5293–5306 (2020)
40. W. Wu, Q. Shen, K. Aldubaikhy, N. Cheng, N. Zhang, X. Shen, Enhance the edge with beamforming: performance analysis of beamforming-enabled WLAN, in *Proceedings of the IEEE WiOpt*, 2018
41. W. Wu, Q. Shen, M. Wang, X. Shen, Performance analysis of IEEE 802.11.ad downlink hybrid beamforming, in *Proceedings of the IEEE ICC*, 2017
42. P. Yang, N. Zhang, S. Zhang, L. Yu, J. Zhang, X. Shen, Content popularity prediction towards location-aware mobile edge caching. IEEE Trans. Multimedia **21**(4), 915–929 (2019)
43. H. Wu, J. Chen, W. Xu, N. Cheng, W. Shi, L. Wang, X. Shen, Delay-minimized edge caching in heterogeneous vehicular networks: a matching-based approach. IEEE Trans. Wireless Commun. **19**(10), 6409–6424 (2020)
44. H. Wu, F. Lyu, C. Zhou, J. Chen, L. Wang, X. Shen, Optimal UAV caching and trajectory in aerial-assisted vehicular networks: a learning-based approach. IEEE J. Sel. Areas Commun. **38**(12), 2783–2797 (2020)
45. W. Wu, N. Zhang, N. Cheng, Y. Tang, K. Aldubaikhy, X. Shen, Beef up mmWave dense cellular networks with D2D-assisted cooperative edge caching. IEEE Trans. Veh. Technol. **68**(4), 3890–3904 (2019)

Chapter 2
Literature Review of mmWave Networks

2.1 Characteristics of mmWave Communication

In order to efficiently utilize mmWave frequency bands, the first and foremost
thing is to understand the mmWave communication characteristics. The mmWave
communication has several unique characteristics as compared to the traditional
microwave communication, including large bandwidth, huge path loss, and sparse
channel, directional antenna, and blockage effect. These propagation characteristics
pose new challenges for establishing reliable connections at mmWave frequency
bands. In the following, these propagation characteristics are introduced in detail.

2.1.1 Large Bandwidth

The first characteristic of mmWave communication is its large bandwidth. The
mmWave frequency band range from 30 GHz to 300 GHz has attracted much
attention from both industry and academia. This is because of the vast amount of
unexplored spectrum in the mmWave frequency band [1]. In specific, mmWave
frequency band possesses more than 20 GHz available spectrum, which consists
of 1.4 GHz bandwidth at the 39 GHz band, 2.1 GHz bandwidth at the 37 GHz and
42 GHz bands, 7 GHz bandwidth at the 60 GHz band, and more than 10 GHz at
the E band.[1] The distribution of available spectrum is depicted in Fig. 2.1. The
amount of available spectrum is 20 times more than that allocated to today's Wi-
Fi and cellular networks. Moreover, if we turn to higher-frequency bands (i.e.,

---

[1] The frequency bands at 71–76 GHz, 81–86 GHz, and 92–95 GHz are collectively referred to as
E Band which are allocated for ultrahigh-speed data communications by federal communications
commission (FCC) in 2003.

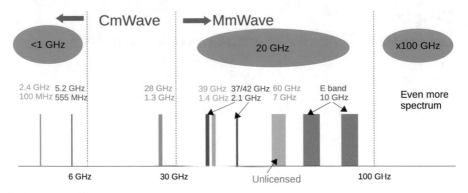

Fig. 2.1 The distribution of available spectrum at the mmWave frequency band

higher than 100 GHz), more than 100 GHz unlicensed bandwidth is available for mmWave communication. Those bands are expected to be utilized in advanced wireless networks, such as the fifth-generation (5G) and future 6G networks, to facilitate extremely high-speed data transmission [2–5].

The large bandwidth contributes to high-speed data transmission. For example, the peak physical rate in the current 802.11ad standard is up to 7 Gbps with a bandwidth of 2.16 GHz [6]. Moreover, in practical systems, experimental results show that the commercial 802.11ad devices are capable of supporting a data rate up to 4.62 Gbps, while the state-of-the-art Wi-Fi, i.e., 802.11ac, can only support a data rate of 866.7 Mbps. The data rate of mmWave networks is much higher than that of existing Wi-Fi systems.

2.1.2 Huge Path Loss

The second characteristic is the significant path loss. We show huge path loss at mmWave communication via simple mathematical calculation in the following.

We consider the ideal *free space transmission* case. In this case, the received power can be characterized by the Friis law formula [7], i.e.,

$$P_r = P_t G_t G_r \left(\frac{\lambda}{4\pi} \right)^2 r^{-n}, \tag{2.1}$$

where P_t represents the transmit power, r is the distance between the transmitter and the receiver, λ is the wavelength of the carrier frequency, and n denotes the path loss exponent. Here, G_t and G_r represent the directional antenna gains of the transmitter and the receiver, respectively.

Based on the above equation, we find that the wavelength of the carrier frequency determines the received power. It is known that mmWave signal wavelength is

Fig. 2.2 Antenna aperture comparison between mmWave communication and microwave communication. A mmWave antenna element captures less RF energy due to a smaller aperture size than a microwave antenna element

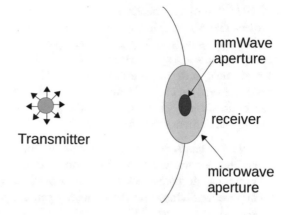

much smaller than microwave signal wavelength. Consequently, mmWave antenna elements have a smaller aperture than microwave antenna elements, such that less RF power can be captured by mmWave antenna elements, which is shown in Fig. 2.2. Therefore, the path loss at the mmWave frequency band is much higher than that at the microwave frequency band. If we consider the unlicensed 60 GHz band, according to the above equation, mmWave communication would suffer from 28 dB extra path loss as compared to that microwave communication at the 2.4 GHz band. Moreover, mmWave communication suffers from higher noise power. This is because mmWave communication has a larger bandwidth than microwave communication. Taking mmWave communication at the 60 GHz band as an example, it possesses 2.16 GHz bandwidth for one channel, which gives 17 dB extra noise power than microwave communication that only possesses 40 MHz bandwidth for one channel. Overall, taking both the received power and noise power into consideration, mmWave communication suffers from severe SNR loss (around 45 dB) compared with the traditional microwave communication.

In addition to ideal free space transmission, *rain attenuation* and *atmospheric absorption* introduce extra path loss for mmWave communication. Rain attenuation is around 7 dB/km in heavy rain scenarios, and atmospheric absorption is about 1 dB/km, based on extensive measurement results at the 28 GHz band [8]. For long-distance mmWave communication, such as wireless backhaul applications, the impact of rain attenuation and atmospheric absorption on the path loss cannot be neglected. However, in urban mobile scenarios, the impact of rain attenuation and atmospheric absorption may be limited. The reason is that the typical radius of mmWave networks is about a few hundred meters. It is worth noting that rain attenuation and atmospheric absorption vary with respect to different frequency bands. For more details, one can refer to detailed in-field measurement results in [9].

With the above knowledge on the path loss, some *path loss models* are constructed. Extensive efforts have been devoted to constructing accurate and practical path loss models in various network environments at different frequency bands, especially those potential frequency bands, such as 28 GHz, 38 GHz, 60 GHz, and

82 GHz. Based on abundant measurement data, the 3rd Generation Partnership Project (3GPP) studying groups provide path loss models for both line-of-sight (LOS) and non-line-of-sight (NLOS) paths. Other measurement campaigns are performed in some specific scenarios, such as airport and high-density stadium [10]. For example, at the unlicensed 60 GHz band, 802.11 consortium performs extensive measurements on channels, and the derived path loss models in various scenarios can be found in [11]. Specifically, the path loss models are categorized based on indoor and outdoor scenarios.

- *Path loss model in indoor scenarios*: At the initial deployment stage, mmWave communication may focus on the indoor scenario. The 802.11ad studying group has devoted much efforts on investigating path loss models for different applications. Based on these efforts, generally, the path loss model in indoor scenarios is given by [11]

$$PL(dB) = A + 20\log_{10}(f) + 10n\log_{10}(d) + \xi, \qquad (2.2)$$

where A and n are parameters based on the specific scenario, f is the carrier frequency (in GHz), d is the distance between the transmitter and the receiver (in meters), and $\xi \in \mathbb{N}(0, \sigma^2)$ is a log normal term accounting for shadowing effect. The detailed parameters in different scenarios are listed in Table 2.1.
- *Path loss model in outdoor scenarios*: In outdoor scenarios, a comprehensive investigation is conducted at the 28 GHz and 73 GHz bands. These two frequency bands are potentially allocated for the fifth-generation 5G cellular networks. Based on experimental measurements in New York City, an empirical path loss model via a linear fitting method is given by

$$PL(dB) = A + 10n\log_{10}(d) + \xi, \qquad (2.3)$$

where A and n are parameters based on the specific scenario and ξ is a log normal term accounting for shadowing effect. Parameters in the above path loss model at different frequency bands are listed in Table 2.2.

Table 2.1 Summary of path loss model parameters in different indoor scenarios. Here, STA means station, LOS is line-of-sight, and NLOS is none-line-of-sight

Scenario	A (dB)	n	Standard deviation (dB)
Conference room with STA-STA LOS path	32.5	2.0	0
Conference room with STA-STA NLOS path	51.5	0.6	3.3
Conference room with STA-AP LOS path	32.5	2.0	0
Conference room with STA-AP NLOS path	45.5	1.4	3.0
Living room with LOS path	32.5	2.0	0
Living room with NLOS path	44.7	1.5	3.4
Cubicle environment with LOS path	32.5	2.0	0
Cubicle environment with NLOS path	44.2	1.8	1.5

Table 2.2 Summary of path
loss model parameters for
outdoor scenarios at different
frequency bands

Parameter	28 GHz	73 GHz
A	72	86.6
n	2.92	2.45
ξ	8.7	8.0

2.1.3   Sparse Channel

The third characteristic of mmWave communication is channel sparsity. The diffraction is very rich in microwave communication at lower-frequency bands which possess dozens of paths in the channel. The channel in mmWave communication is relatively sparse, and there are only a few clustered paths in the channel. In the following, we show the number of paths and the performance of these paths in practical scenarios:

- *Number of paths*: Commonly, the clustered paths are made up of one dominant LOS path and a few NLOS paths. These NLOS paths are usually generated by some strong reflectors, such as human bodies and building materials. Recent experiments and measurements show that the mmWave channel is very sparse [12]. For example, in office, corridor, and conference room scenarios, there are less than five paths on average at the unlicensed 60 GHz band. At the 28 GHz band, similar results have been obtained. In addition, another study in [13] shows that only averagely 2.4 clustered paths exist in mmWave channel. As a typical example shown in Fig. 2.3, only two clustered paths exist between the transmitter and receiver. One is a LOS path, and the other is a NLOS path which is caused by the strong metallic reflector.
- *Performance of paths*: Extensive experiments are conducted to study the performance of the paths, especially NLOS paths. It is expected that the NLOS path suffers extra path loss than the LOS path due to the reflection and a longer transmission distance. Measurement results based on a ray-tracing method in an office environment show that the first-order reflection NLOS path suffers about 5–10 dB SNR loss compared with the LOS path. The SNR of the second-order reflection NLOS path is approximately 10–20 dB lower than that of the LOS path [14].

Although channel sparsity can reduce network coverage, it also brings some *benefits*. For example, channel sparsity feature can be utilized to address the channel estimation problem in mmWave communications. A large antenna array is exploited to form narrow beams with high antenna gains to counteract the path loss at mmWave frequency bands. The adoption of the large antenna array incurs a formidable task for channel estimation since the dimension of channel matrix is large, which results in a significant channel estimation overhead. As such, channel estimation becomes a bottleneck for mmWave communication to achieve high

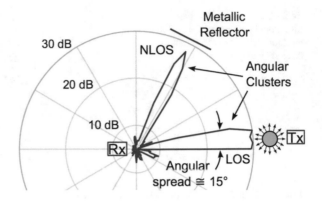

Fig. 2.3 The LOS and NLOS paths in mmWave communication. The NLOS path is caused by reflection

throughput. To solve the issue, channel sparsity in mmWave communication can be harnessed. Some compressive sensing-based algorithms are designed to alleviate the cost and complexity of channel estimation via [15]. The authors in [16] proposed a fast beam tracking algorithm via exploiting channel sparsity feature at mmWave frequency bands.

2.1.4   Directional Antennas

Another characteristic in mmWave communications is the directional antennas. Unlike the traditional microwave communication systems that adopt omni-directional antennas, mmWave communications need to adopt directional antennas to compensate huge path loss. To achieve sufficient antenna gain, the beamwidth is usually very narrow. For example, a 50×50 phased antenna array can generate a narrow beam with a beamwidth of $3°$, which can provide a directional antenna gain up to 36 dB. The high-directionality characteristic is one of the key distinctions between mmWave communication and traditional microwave communication. High directionality results in many challenges in mmWave communications, such as beam alignment and medium access control performance, which are detailed in Chaps. 3 and 4.

2.1.5   Blockage Effect

The last characteristic is the blockage. Unlike microwave signals, mmWave signals suffer from *high penetration loss*. This is because the mmWave signal wavelength is usually much smaller than the sizes of objects in the environment. For example, a brick can attenuate mmWave signals as much as 40–80 dB [17, 18]. The human body may result in an attenuation of 20–35 dB. Foliage also causes a significant penetration loss, which poses significant challenges for outdoor mmWave communications. The experience high penetration loss in mmWave communication is in stark contrast with microwave communication [19].

The high penetration loss makes *blockage* as a thorny issue in mmWave communications. There are several forms of common blockage in mmWave communications, and some are listed as follows: (1) hand blockage, i.e., mmWave signals are blocked by human hands, which may occur with a high probability, as mobile phones or other portable digital devices are usually held in human hands, and (2) self-body blockage, which is caused by human gesture and body rotation. This blockage form is common in indoor scenarios. These blockage forms do not occur in traditional microwave communications. In practical scenarios, various kinds of blockage forms need to be considered and addressed. We can claim that blockage can be considered as one of the main differences between the traditional microwave communication and mmWave communication [20].

The *blockage effect* varies in different scenarios. Some studies investigate the blockage effect in different scenarios. For indoor scenarios, blockage is mainly caused by human movements. Experiment results in [21] show that link outage caused by blockage increases to 22% when there are 11–15 persons in the room, and the link outage decreases to only 1%–2% when there are 1–5 persons in the room. In addition, the duration of blockage effect may last several hundred milliseconds. The latency incurs significant data transmission delay. For outdoor scenarios, the LOS link can be easily blocked by buildings and other static objects especially in a dense environment. This blockage shows that the coverage of mmWave network can be severely affected by nearby environments. Besides, if blockage occurs, the received SNR would drop significantly, rendering link outage and great degradation of network throughput.

In the literature, some *anti-blockage* schemes are proposed to alleviate the blockage effect. First, to understand blockage effect better, random shape theory and geographic information are applied to derive some models to analyze blockage effect [22]. Then, to solve the blockage issue of the dominant LOS link, strong NLOS paths are exploited to provide a reliable connection when blockage occurs [14]. For example, measurement results in New York City show that NLOS paths can be used to expand the network coverage even in a dense urban scenario [23]. By predicting the performance of multiple beams, "BeamSpy" is designed to immediately choose the best alternative beam to re-establish the communication link once blockage occurs [12].

2.2   Beamforming Technology

Beamforming is the key enabling technology in mmWave communication to overcome huge path loss. There are several state-of-the-art beamforming technologies proposed in the literature. We first introduce digital beamforming and analog beamforming technologies in Sects. 2.2.1 and 2.2.2, respectively. Then, we introduce a hybrid version, i.e., hybrid beamforming, for supporting multiuser transmission in Sect. 2.2.3.

2.2.1 Digital Beamforming

Digital beamforming is usually adopted in traditional microwave communication at the below 6 GHz frequency band, which is performed by digital baseband processing components, such as digital signal processing (DSP) units. However, applying existing digital beamforming in mmWave communication systems faces multiple challenges: (1) Excessive energy consumption—In the existing digital beamforming architecture, each antenna element needs to connect with a radio-frequency (RF) chain and an analog-to-digital conversion (ADC)/digital-to-analog conversion (DAC). Both RF chain and ADC/DAC consume a large amount of energy. Moreover, mmWave communication systems usually adopt a large number of antenna elements (e.g., 1024), and thus a large number of RF chains and ADCs/DACs should be installed for digital beamforming, resulting in excessive energy consumption. (2) Prohibitive hardware complexity—It is difficult to place so many RF chains and ADCs/DACs into a small chip fabrication. Therefore, taking both energy consumption and hardware constraints into consideration, digital beamforming may be unsuitable in mmWave systems.

2.2.2 Analog Beamforming

Another beamforming technology is digital beamforming, which is a de facto approach in mmWave systems to reduce complexity. Analog beamforming has been adopted in many ratified mmWave communication standards, such as 802.11ad and WirelessHD. As shown in Fig. 2.4, analog beamforming only employs one

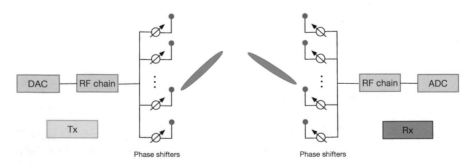

Fig. 2.4 The analog beamforming architecture in mmWave communication. Analog circuit adjusts the beam direction via shifting phases of each antenna element

RF chain and one DAC/ADC at the transmitter or the receiver. Specifically, analog phase shifters are utilized to implement analog beamforming. Giving each antenna a phase shift, i.e., a phase weight vector at the antennas, analog beamforming can be generated. For low implementation complexity consideration, the phase weight vectors can be designed offline and stored in a codebook. A codebook is a matrix, in which each column vector indicates a phase weight vector for the antenna array and can be used to form a specific beam pattern. Multiple codebook-based analog beamforming schemes are proposed to form various beam patterns in the literature. For example, a codebook based on discrete Fourier transform (DFT) is proposed [24]. This codebook has been widely adopted to generate uniform antenna gain for different beam directions in recent studies in mmWave networks.

2.2.3 Hybrid Beamforming

The *motivation* of hybrid beamforming is due to the following factors. Although analog beamforming can offer high-speed data transmission, it can only support one data stream at a time. To enable multiuser transmission in mmWave networks, extensive efforts are devoted recently. Multiuser transmission can significantly enhance spatial reuse and increase data rate in mmWave networks. Taking the current 802.11ac standard at lower-frequency bands as an example, 802.11ac can support up to four users in the multiuser transmission mode. The multiuser transmission schemes in lower frequency bands are implemented by digital beamforming which can effectively mitigate the interference among users to enhance data rate. As we mentioned in Sect. 2.2.1, the design of digital beamforming faces many challenges in mmWave systems, such as high hardware complexity, huge channel estimation overhead, and excessive energy consumption. These challenges make digital beamforming unsuitable for mmWave systems. It is desired to develop a tailored energy-efficient yet low-complexity solution for mmWave communication.

A novel solution, namely, *hybrid beamforming that integrates analog beamforming and digital beamforming*, is proposed to enable multiuser transmission in mmWave systems [25]. In specific, the analog beamforming part controls the signal phase at each antenna element to provide sufficient directional antenna gain, and the digital beamforming part focuses on mitigating interference among users. A typical example is shown in Fig. 2.5. The hybrid beamforming scheme requires multiple RF chains to support multiple data streams. Existing works show that hybrid beamforming can achieve close-to-optimal performance as compared to fully digital beamforming while significantly reducing complexity [26]. As such, if hybrid beamforming could be successfully applied for multiuser downlink transmission, data rate in mmWave networks can be significantly improved.

A collection of works investigate hybrid beamforming performance in different mmWave networks. The first hybrid beamforming algorithm for mmWave networks is proposed by Alkhateeb et al. in [27], which leverages the channel state information feedback from users to design the baseband digital beamforming at the AP.

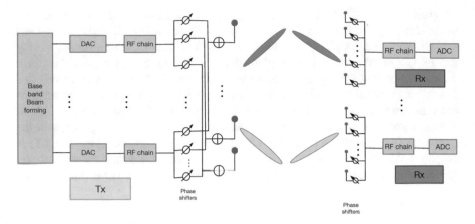

Fig. 2.5 An illustrative example of hybrid beamforming in mmWave communication

Another important work utilized the channel sparsity to design a low-complexity hybrid beamforming algorithm in [28]. An extended work investigated efficient hybrid beamforming design in mmWave cellular networks [29]. Aldubaikhy et al. studied the user selection issue in the hybrid beamforming algorithm, and then they proposed a novel low-complexity hybrid beamforming scheme which incorporates user selection to enhance the uplink performance in mmWave systems [30]. Hybrid beamforming-based mmWave communication can also be applied for supporting data transmission services for high-mobility trains [31].

2.3 Beamforming Training Protocol in IEEE 802.11ad

As mentioned before, 802.11ad is the first ratified WLAN standard that operates in the unlicensed 60 GHz band. The 802.11ad has achieved great success in mmWave WLANs and is going to witness an emerging great market of mmWave WLAN devices in the future. In 802.11ad, beamforming plays a pivotal role to establish reliable mmWave connection between the transmitter and receiver. The transmitter and receiver must scan their entire beam space to find the best transmit and receive beams. Beamforming training is such a procedure to find the best transmit and receive beams.

In this section, we give a detailed introduction for the beamforming training protocol in the celebrated 802.11ad standard. Specifically, we first give an overview of the protocol in Sect. 2.3.1. For two essential phases in the protocol, i.e., sector level sweep (SLS) and beam refinement protocol (BRP), we present not only their basic ideas but also corresponding detailed operations in the standard in Sects. 2.3.2 and 2.3.3, respectively. Finally, beam searching complexity analysis of the protocol is presented in Sect. 2.3.4.

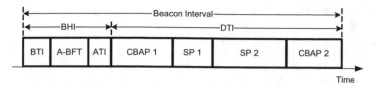

Fig. 2.6 Beacon interval structure in 802.11ad

2.3.1   Overview of 802.11ad Beamforming Training Protocol

There are two operation modes in 802.11ad. One is the directional multi-gigabit (DMG) mode, and the other is the non-DMG mode. The DMG mode in 802.11ad is to adopt beamforming technology in mmWave communication to provide high-speed data transmission. In the following, we focus on the protocol behavior in the DMG mode.

We first introduce the frame structure in 802.11ad. In 802.11ad, beacon interval (BI) is the basic time frame [32, 33]. As shown in Fig. 2.6, a BI is made up of two stages: (1) beacon header interval (BHI) stage and (2) data transmission interval (DTI) stage. The BHI stage consists of beacon transmission interval (BTI), association beamforming training (A-BFT), and announcement transmission intervals (ATI) stages. The DTI stage includes multiple service periods (SPs) and contention-based access periods (CBAPs), which are allocated for data transmission with different purposes. Specifically, SPs are for scheduled access periods, and CBAPs are for distributed channel access periods.

Based on the BI structure, the working flow of the beamforming training protocol is presented. Specifically, the protocol is made up of three phases: sector level sweep (SLS), beam refinement protocol (BRP), and beam tracking (BT). The functionalities of these phases are detailed as follows:

- *SLS phase*: This phase is performed in the BTI and A-BFT stages. Both the transmitter and receiver sweep all the sectors to obtain the optimal transmit and receive sectors (i.e., *coarse-grained* or wide beams). A sector includes several fine-grained beams. As the example shown in Fig. 2.7, there are eight sectors in the entire beam space, and each sector contains six fine-grained beams.
- *BRP phase*: This phase is performed in the DTI stage. Once the optimal transmit and receive sectors are identified, both the transmitter and receiver search all the fine-grained beams contained in the identified sectors to find the *fine-grained* transmit-receive beams (i.e., narrow beams).
- *BT phase*: BT is an optional phase which is employed during the DTI stage to adapt to channel change. For example, in high-mobility scenarios, such as vehicular networks, the beamforming training should be invoked continuously, which is to track beams between the transmitter and receiver.

This chapter focuses on the two compulsory phases, i.e., SLS and BRP. Their detailed descriptions and operations are given in the following.

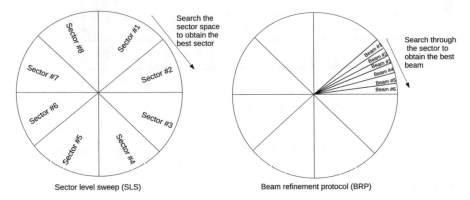

Fig. 2.7 Coarse-grained beams (i.e., sectors) in the SLS phase and fine-grained beams in the BRP phase

2.3.2 Sector Level Sweep

The *goal* of the SLS phase is to obtain the *optimal transmit-receive sector pair*, i.e., coarse-grained beam pair. Specifically, the SLS phase consists of the following two steps:

1. The transmitter adopts the directional transmission mode, i.e., all the sectors in the sector space are selected in a sequential manner, and the receiver keeps the omni-directional mode. As such, the receiver can obtain the *optimal transmit sector* identification (ID) via comparing the received signal strength of all the sectors.
2. The second step is to perform the first step in an opposite way. The transmitter keeps the omni-directional mode, while the receiver adopts the directional mode. Similarly, the *optimal receiver sector* ID can be obtained.

As such, the optimal sector pair between the transmitter and receiver can be obtained. The basic idea of the SLS phase is given in Fig. 2.8. This process is compulsory which must be performed in the BTI and A-BFT phases. The BTI phase is for the first step, and the A-BFT phase is for the second step. Note that beamforming training in the SLS phase should be judiciously designed since the length of A-BFT phase is limited.

Next, we show the *detailed operations* in the 802.11ad standard to implement the SLS phase. We consider an SLS example between an access point (AP) and one station (STA), i.e., user, as shown in Fig. 2.9. Here, the AP is the transmitter, and the STA is the receiver. The detailed procedure consists of four subphases, i.e., initial sector sweep (ISS), responder sector sweep (RSS), sector sweep feedback, and sector sweep acknowledgment (ACK) subphases. These subphases operate as follows:

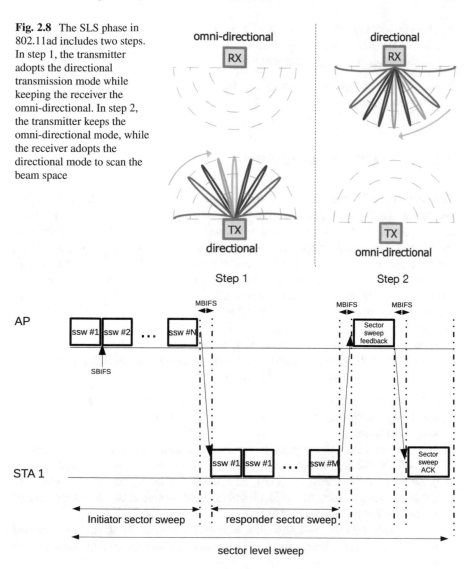

Fig. 2.8 The SLS phase in 802.11ad includes two steps. In step 1, the transmitter adopts the directional transmission mode while keeping the receiver the omni-directional. In step 2, the transmitter keeps the omni-directional mode, while the receiver adopts the directional mode to scan the beam space

Fig. 2.9 Operations between an AP and an STA in the SLS phase

- In the ISS subphase, multiple sector sweep (SSW) frames are transmitted by the AP using different sectors. During the transmission, the STA keeps the omni-directional mode to receive the SSW frames. The optimal transmit sector ID at the AP is identified by comparing received signal strength at the STA.
- In the RSS subphase, the process is reversed. The STA scans its sector space via transmitting SSW frames to the AP, while the AP adopts the quasi-omni-directional mode. Similar to that in the ISS subphase, the AP can identify the optimal receive sector ID at the STA according to the received signal strength.

- The best receive and transmit quasi-omni sector IDs are exchanged between the AP and the STA via sending a sector sweep feedback frame and a sector sweep acknowledgment frame in the last two subphases.

In this way, the SLS phase is completed, and the optimal transmit-receive sector pair is identified.

2.3.3 Beam Refinement Protocol

The SLS phase is followed by the BRP phase which is to refine the beams. The *goal* of the BRP phase is that both the transmitter and receiver search all the fine-grained beams within the identified transmit-receive sector pair to find the best transmit-receive fine-grained beam. The BRP phase consists of four subphases, i.e., BRP setup, multiple sector ID (MID), beam combining (BC), and BRP transaction subphases. Due to imperfect quasi-omni-directional antennas in practical mmWave systems, the MID and BC subphases are used to identify better initial antenna weight vectors via iterative beam refinement. The number of iterations to finish MID and BC subphases is denoted by N_{BRP}.

Next, we present the *operations* in the BRP phase via the following simple example. The BRP phase is performed between an AP and an STA, as shown in Fig. 2.10. The procedure is detailed as follows:

1. *BRP setup subphase*: The STA and the AP first exchange beam refinement capability information and request the execution of the other BRP subphases via some BRP frames.
2. *MID subphase*: In this subphase, the transmitter firstly scans all the fine-grained beams in the identified optimal transmit sector to obtain several candidate fine-grained transmit beams. Secondly, in a similar manner, the receiver scans all the fine-grained beams in the identified optimal receive sector to obtain several candidate fine-grained receive beams.
3. *BC subphase*: Both the transmitter and receiver scan candidate fine-grained transmit and receive beams to identify the optimal transmit-receive fine-grained beam pair.
4. *BRP transaction subphase*: The information between the AP and the STA is exchanged to report the optimal transmit-receive beam pair.

Note that BRP is performed in the DTI stage which is different from the SLS phase in the BTI and A-BFT stages. One can refer to the 802.11ad standard for the detailed operations.

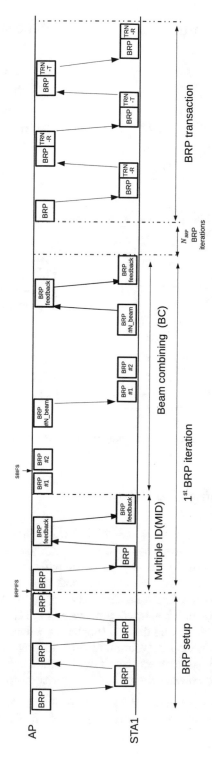

Fig. 2.10 Operations in the BRP phase between an AP and an STA, including BRP setup, MID, BC, and BRP transaction subphases

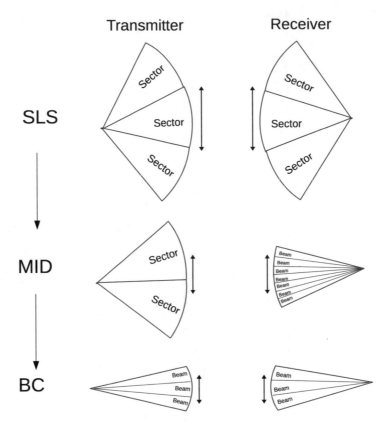

Fig. 2.11 Beam searching in 802.11ad beamforming training protocol, including SLS, MID, and BC phases

2.3.4 Beam Searching Complexity

For the beamforming training protocol in 802.11ad, the beam searching complexity analysis of the protocol is given as follows. To analyze the complexity, the illustration of the beamforming training protocol with the SLS and two BRP subphases is given in Fig. 2.11. We assume that both the transmitter and receiver adopt the same beam patterns. The transmitter is equipped with Q sectors. There are $B > Q$ narrow beams in the entire beam space [19, 34]. In the protocol, the SLS phase, the MID subphase, and the BC subphase conduct the beam searching operations. The reason that the BRP setup and BRP transaction subphases are not considered is because there is no beam searching operation in these phases.

The beam searching complexity analysis in each phases is given as follows:

- *Beam searching complexity in the SLS phase*: To find the optimal transmit and receive sectors, the transmitter and receiver search all the Q sectors in their sector space via an exhaustive searching algorithm. The corresponding searching complexity is Q^2.

- *Beam searching complexity in the MID subphase*: Let $\delta \geq 2$ denote the number of candidate transmit quasi-omni sectors. All the fine-grained beams in the optimal receive quasi-omni sector are scanned to obtain optional candidate transmit and receive fine-grained beams. The corresponding searching complexity is $\delta B/Q$.
- *Beam searching complexity in the BC subphase*: Let $\gamma \leq 7$ denote the number of candidate fine-grained beams. All γ beams are scanned at both transmitter and receiver to find the best transmit and receive fine-grained beams via an exhaustive method. The corresponding searching complexity is γ^2.

Overall, taking the above three phases into consideration, the searching complexity of the beamforming training protocol in 802.11ad is given by

$$BF_\tau = Q^2 + \frac{\delta B}{Q} + \gamma^2. \tag{2.4}$$

The above equation indicates that the beam searching complexity of the protocol is $O(Q^2)$. Compared with the beam searching complexity via an exhaustive searching method whose complexity is $O(B^2)$, the beam searching complexity of the beamforming training protocol in 802.11ad is greatly reduced.

From the results, we show that beam searching complexity increases quadratically with the number of beams. We know that narrow beams would be adopted in mmWave communication to provide sufficient directional antenna gain for high-speed data transmission. The adoption of narrow beams would incur a significant high beam searching complexity. Such high beam searching complexity renders a long access delay. For a narrow beam with a beamwidth of 2.8°, it takes approximately 4 seconds for finishing the beamforming training procedure [35]. In high-mobility scenarios, such as vehicular networks, frequent beamforming training is required due to beam misalignment caused by user mobility, which degrades mmWave network performance. This motivates our work on designing low-latency beam alignment schemes in Chap. 3.

2.4 Multi-armed Bandit Theory

In this section, we introduce some *basic knowledge* on multi-armed bandit (MAB) theory. It is worth noting that the MAB method is one of reinforcement learning methods. Recently, traditional reinforcement learning methods, such as deep Q networks, have been widely applied to solve networking problems in the wireless domain, including mobile crowdsourcing [36], service migration in vehicular networks [37], network slicing in cellular networks [38], task offloading in industrial IoT networks [39], resource scaling in virtual networks [40], and content caching in edge networks [41]. Compared with the traditional reinforcement learning methods

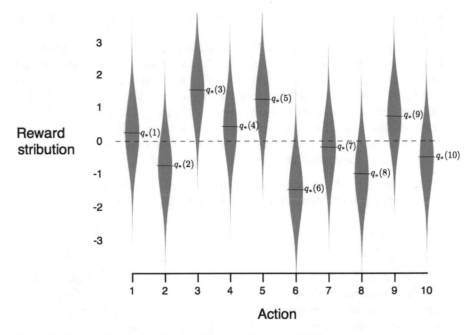

Fig. 2.12 Reward distributions for ten different actions in an MAB problem

with a tuple of three elements, i.e., state, action, and reward, the MAB method only has two elements, i.e., action and reward, which lack the state. Such difference results in different applications of traditional reinforcement learning methods and the MAB method.

The MAB theory *originates* from an *exploration* and *exploitation* dilemma. The exploration is to make the best decision given current information. The exploitation is to gather more information in order to make better decisions. We often face this dilemma in our daily life. Taking video games as an example, we often have two choices. One option is to play the move that you believe is the best. This option is exploitation. Another option is to play an experimental move. This option is exploration. Another example is the online banner adverts placement for business groups. Showing the most successful advert to users based on their historical records is exploitation. Showing a different advert for users is exploration. With such exploration and exploitation dilemma, the optimal long-term strategy involves short-term sacrifices. On the one hand, gathering enough information via exploration to make the best overall decisions is required. On the other hand, frequently, exploration may waste the constrained resources. Hence, it is a challenging issue to balance the exploration and exploitation.

The MAB theory is used to address such exploration and exploitation dilemma. Its name originates from the old-fashioned one-armed bandit machine game in the casino. In this game, in front of a gambler, there are multiple one-armed bandit machines. The gambler pulls an arm of a bandit machine at each time, and then he gets a reward from the bandit machine. The goal of the gambler is to maximize his long-term benefit by selecting the arm of bandit machines in a sequential manner. This problem is referred to as the *stochastic MAB problem*. Assume that there are N bandit machines. Each bandit machine has an independent reward distribution, and the reward distribution is unknown to the gambler. As the example shown in Fig. 2.12, there are ten candidate actions, and the reward distribution of each action is independent. As such, the obtained reward at each time is a sample from the reward distribution. Mathematically, stochastic MAB problem can be modeled by a tuple with two elements, i.e., $(\mathcal{A}, \mathcal{R})$. Here, $\mathcal{A}$ represents the set of *actions*, which are referred to as *arms*, and $\mathcal{R}$ represents the set of unknown reward distributions. Let t denote the index of time slots, i.e., we have $t = 1, 2, \ldots, T$. At each time slot, action a_t is selected by an MAB algorithm, and then a reward r_t is obtained from the environment. The objective of the MAB algorithm is to find a policy to maximize the cumulative reward. In the bandit theory, people usually adopt *cumulative regret* to evaluate the performance of a bandit learning algorithm. Assume that there exists the optimal algorithm which knows the optimal action *a priori* and always selects the optimal action. The cumulative regret is defined as the gap between the optimal algorithm and the adopted bandit learning algorithm. Based on the above definition, maximizing the cumulative reward is equivalent to minimizing the cumulative regret.

In the literature, a number of solutions for MAB problems have been proposed in different settings. For a standard stochastic MAB problem, Lai and Robbins first studied the fundamental regret lower bound in [42]. In the following, Auer et al. proposed the celebrated upper confidence bound (UCB) algorithm to achieve the lower bound [43]. Based on the core idea of UCB, many variants of UCB algorithms have been developed for different problems, including (1) Bayesian bandit problem [44, 45], in which some prior knowledge is leveraged to speed up the learning process; (2) combinatorial bandit problem [46], whose objective is to select the optimal subset of actions instead of selecting a single action; (3) volatile bandit algorithm [47], in which the set of action varies; (4) correlated bandit problem [48], in which actions are correlated with each other; and (5) multi-agent bandit problem [49], in which multiple agents independently learn a joint action set. Interested readers are referred to a tutorial survey in [50] and the references therein. In Chap. 3, a correlated MAB problem is investigated in the context of beam alignment in mmWave networks [51].

2.5 Summary

In this chapter, we have surveyed the main characteristics of mmWave communication and the state-of-the-art beamforming training techniques for establishing mmWave connections. In addition, we have presented the beamforming training protocol in the celebrated 802.11ad standard in detail. Also, we have introduced some basic knowledge on the MAB theory for its potential application in the next chapter.

References

1. J.G. Andrews, S. Buzzi, W. Choi, S.V. Hanly, A. Lozano, A.C. Soong, J.C. Zhang, What will 5G be? IEEE J. Sel. Areas Commun. **32**(6), 1065–1082 (2014)
2. X. You et al., Towards 6G wireless communication networks: Vision, enabling technologies, and new paradigm shifts. Sci. China Inf. Sci. **64**(1), 1–74 (2021)
3. X. Shen, J. Gao, W. Wu, K. Lyu, M. Li, W. Zhuang, X. Li, J. Rao, AI-assisted network-slicing based next-generation wireless networks. IEEE Open J. Veh. Technol. **1**(1), 45–66 (2020)
4. N. Zhang, P. Yang, J. Ren, D. Chen, L. Yu, X. Shen, Synergy of big data and 5G wireless networks: Opportunities, approaches, and challenges. IEEE Wireless Commun. **25**(1), 12–18 (2018)
5. N. Zhang, N. Cheng, A.T. Gamage, K. Zhang, J.W. Mark, X. Shen, Cloud assisted hetnets toward 5G wireless networks. IEEE Commun. Mag. **53**(6), 59–65 (2015)
6. S. Sur, I. Pefkianakis, X. Zhang, K.H. Kim, WiFi-assisted 60 GHz wireless networks, in *Proc. ACM MOBICOM* (2017), pp. 28–41
7. H.T. Friis, A note on a simple transmission formula. Proc. IRE **34**(5), 254–256 (1946)
8. M. Xiao, S. Mumtaz, Y. Huang, L. Dai, Y. Li, M. Matthaiou, G.K. Karagiannidis, E. Björnson, K. Yang, I. Chih-Lin, Millimeter wave communications for future mobile networks. IEEE J. Sel. Areas Commun. **35**(9), 1909–1935 (2017)
9. E-Band Communications, E-band technology. [Online]. Available: http://www.e-band.com/index.php?id=86
10. mmMAGIC, Measurement campaigns and initial channel models for preferred suitable frequency ranges (2016). [Online]. Available: https://5g-mmmagic.eu/results/deliverable
11. V.E. A. Maltsev, Channel Models for 60 GHz WLAN Systems. Doc.:IEEE 802.11-09/0334r8, vol. 3
12. S. Sur, X. Zhang, P. Ramanathan, R. Chandra, BeamSpy: Enabling robust 60 GHz links under blockage, in *Proc. USENIX NSDI* (2016), pp. 193–206
13. M. Samimi, K. Wang, Y. Azar, G.N. Wong, R. Mayzus, H. Zhao, J. K. Schulz, S. Sun, F. Gutierrez, T.S. Rappaport, 28 GHz angle of arrival and angle of departure analysis for outdoor cellular communications using steerable beam antennas in New York City, in *Proc. IEEE VTC Spring* (2013), pp. 1–6
14. A. Maltsev, R. Maslennikov, A. Sevastyanov, A. Khoryaev, A. Lomayev, Experimental investigations of 60 GHz WLAN systems in office environment. IEEE J. Sel. Areas Commun. **27**(8), 1488–1499 (2009)
15. D.E. Berraki, S.M. Armour, A.R. Nix, Application of compressive sensing in sparse spatial channel recovery for beamforming in mmwave outdoor systems, in *Proc. IEEE WCNC* (2014), pp. 887–892
16. S.C. Han Yan, D. Cabric, Wideband channel tracking for mmwave mimo system with hybrid beamforming architecture. Proc. IEEE ICASSP, 1–5 (2017)
17. Z. Pi, F. Khan, An introduction to millimeter-wave mobile broadband systems. IEEE Commun. Mag. **49**(6), 101–107 (2011)
18. C.R. Anderson, T.S. Rappaport, In-building wideband partition loss measurements at 2.5 and 60 GHz. IEEE Trans. Wireless Commun. **3**(3), 922–928 (2004)
19. S. Sur, V. Venkateswaran, X. Zhang, P. Ramanathan, 60 GHz indoor networking through flexible beams: A link-level profiling. ACM SIGMETRICS Perform. Eval. Rev. **43**(1), 71–84 (2015)
20. F. Boccardi, R.W. Heath, A. Lozano, T.L. Marzetta, P. Popovski, Five disruptive technology directions for 5G. IEEE Commun. Mag. **52**(2), 74–80 (2014)
21. S. Collonge, G. Zaharia, G.E. Zein, Influence of the human activity on wide-band characteristics of the 60 GHz indoor radio channel. IEEE Trans. Wireless Commun. **3**(6), 2396–2406 (2004)
22. T. Bai, R. Vaze, R.W. Heath, Analysis of blockage effects on urban cellular networks. IEEE Trans. Wireless Commun. **13**(9), 5070–5083 (2014)

23. T.S. Rappaport, S. Sun, R. Mayzus, H. Zhao, Y. Azar, K. Wang, G.N. Wong, J.K. Schulz, M. Samimi, F. Gutierrez, Millimeter wave mobile communications for 5G cellular: It will work! IEEE Access **1**, 335–349 (2013)

24. Y. Shabara, C.E. Koksal, E. Ekici, Linear block coding for efficient beam discovery in millimeter wave communication networks, in *Proc. IEEE INFOCOM* (2018), pp. 2285–2293

25. R. Méndez-Rial, C. Rusu, N. González-Prelcic, A. Alkhateeb, R.W. Heath, Hybrid MIMO architectures for millimeter wave communications: Phase shifters or switches? IEEE Access **4**, 247–267 (2016)

26. F. Sohrabi, W. Yu, Hybrid digital and analog beamforming design for large-scale antenna arrays. IEEE J. Sel. Topics Signal Process. **10**(3), 501–513 (2016)

27. A. Alkhateeb, R.W. Heath, Frequency selective hybrid precoding for limited feedback millimeter wave systems. IEEE Trans. Commun. **64**(5), 1801–1818 (2016)

28. C. Rusu, R. Méndez-Rial, N. González-Prelcicy, R.W. Heath, Low complexity hybrid sparse precoding and combining in millimeter wave MIMO systems, in *Proc. IEEE ICC* (2015), pp. 1340–1345

29. A. Alkhateeb, O. El Ayach, G. Leus, R.W. Heath, Hybrid precoding for millimeter wave cellular systems with partial channel knowledge, in *Proc. IEEE ITA* (2013), pp. 1–5

30. K. Aldubaikhy, W. Wu, Q. Ye, X. Shen, Low-complexity user selection algorithms for multiuser transmissions in mmwave WLANs. IEEE Trans. Wireless Commun. **19**(4), 2397–2410 (2020)

31. M. Gao, B. Ai, Y. Niu, W. Wu, P. Yang, F. Lyu, X. Shen, Efficient hybrid beamforming with anti-blockage design for high-speed railway communications. IEEE Trans. Veh. Technol. **69**(9), 9643–9655 (2020)

32. W. Wu, N. Cheng, N. Zhang, P. Yang, K. Aldubaikhy, X. Shen, Performance analysis and enhancement of beamforming training in 802.11ad. IEEE Trans. Veh. Technol. **69**(5), 5293–5306 (2020)

33. W. Wu, Q. Shen, M. Wang, X. Shen, Performance analysis of IEEE 802.11. ad downlink hybrid beamforming, in *Proc. IEEE ICC* (2017), pp. 1–6

34. O. Abari, H. Hassanieh, M. Rodriguez, D. Katabi, Millimeter wave communications: From point-to-point links to agile network connections, in *Proc. ACM HotNets* (2016), pp. 169–175

35. T. Wei, A. Zhou, X. Zhang, Facilitating robust 60 ghz network deployment by sensing ambient reflectors, in *Proc. USENIX NSDI* (2017), pp. 213–226

36. P. Yang, N. Zhang, S. Zhang, K. Yang, L. Yu, X. Shen, Identifying the most valuable workers in fog-assisted spatial crowdsourcing. IEEE Internet Things J. **4**(5), 1193–1203 (2017)

37. S. Wang, Y. Guo, N. Zhang, P. Yang, A. Zhou, X. Shen, Delay-aware microservice coordination in mobile edge computing: A reinforcement learning approach. IEEE Trans. Mobile Comput. **20**(3), 939–951 (2021)

38. W. Wu, N. Chen, C. Zhou, M. Li, X. Shen, W. Zhuang, X. Li, Dynamic RAN slicing for service-oriented vehicular networks via constrained learning. IEEE J. Sel. Areas Commun. **39**(7), 2076–2089 (2021)

39. W. Wu, P. Yang, W. Zhang, C. Zhou, X. Shen, Accuracy-guaranteed collaborative DNN inference in industrial IoT via deep reinforcement learning. IEEE Trans. Ind. Inf. **17**(7), 4988–4998 (2021)

40. K. Qu, W. Zhuang, X. Shen, X. Li, J. Rao, Dynamic resource scaling for VNF over nonstationary traffic: A learning approach. IEEE Trans. Cogn. Commun. Netw. **7**(2), 648–662 (2021)

41. P. Yang, N. Zhang, S. Zhang, L. Yu, J. Zhang, X. Shen, Content popularity prediction towards location-aware mobile edge caching. IEEE Trans. Multimedia **21**(4), 915–929 (2019)

42. T.L. Lai, H. Robbins, Asymptotically efficient adaptive allocation rules. Adv. Appl. Math. **6**, 4–22 (1985)

43. P. Auer, N. Cesa-Bianchi, P. Fischer, Finite-time analysis of the multiarmed bandit problem. Mach. Learn. **47**(2), 235–256 (2002)

44. P.B. Reverdy, V. Srivastava, N.E. Leonard, Modeling human decision making in generalized Gaussian multiarmed bandits. Proc. IEEE **102**(4), 544–571 (2014)

45. Z. Wang, C. Shen, Small cell transmit power assignment based on correlated bandit learning. IEEE J. Sel. Areas Commun. **35**(5), 1030–1045 (2017)
46. W. Chen, Y. Wang, Y. Yuan, Combinatorial multi-armed bandit: General framework and applications, in *Proc. ICML* (2013), pp. 151–159
47. L. Chen, J. Xu, Z. Lu, Contextual combinatorial multi-armed bandits with volatile arms and submodular reward, in *Proc. NIPS* (2018), pp. 3247–3256
48. C. Shen, R. Zhou, C. Tekin, M. van der Schaar, Generalized global bandit and its application in cellular coverage optimization. IEEE J. Sel. Topics Signal Process. **12**(1), 218–232 (2018)
49. C. Claus, C. Boutilier, The dynamics of reinforcement learning in cooperative multiagent systems. AAAI/IAAI **1998**, 746–752 (1998)
50. S. Bubeck, Cesa-Bianchi et al., Regret analysis of stochastic and nonstochastic multi-armed bandit problems. Found. Trends® Mach. Learn. **5**(1), 1–122 (2012)
51. W. Wu, N. Cheng, N. Zhang, P. Yang, W. Zhuang, X. Shen, Fast mmwave beam alignment via correlated bandit learning. IEEE Trans. Wireless Commun. **18**(12), 5894–5908 (2019)

Chapter 3
Machine Learning-Based Beam Alignment in mmWave Networks

3.1 Introduction

In mmWave communication systems, both the transmitter and receiver use narrow directional beams to make up for the huge attenuation loss [1]. The communication can take place only if the beams of the transmitter and receiver are correctly aligned [2], as shown in Fig. 3.1. Beam alignment (BA) is such a process that identifies the optimal transmit-receive beam pair for maximizing the received signal strength (RSS). Beam misalignment can reduce the link budget remarkably and cut down the throughput from several Gbps to a few hundred Mbps [3]. As a pivotal process in mmWave communications, BA is of great significance to the realization of multi-gigabit wireless transmission. A naive exhaustive search method scans all the combinations of the transmitter and receiver beams to obtain the best beam pair, but this causes a large BA delay. However, a low-latency BA process is essential for practical mmWave systems to adopt real-time applications. Furthermore, in mobile scenarios, user mobility changes the beam direction and thus frequently calls BA, which further aggravates the delay. To accelerate the beam search, 802.11ad protocol decouples the BA process into two steps. Firstly, the transmitter begins with a quasi-omni-directional beam, and the receiver scans the beam space to obtain the best receiver beam. Secondly, the transmitter scans the beam space for the best transmitter beam and keeps the receiver quasi-omni-directional in the meantime. Nevertheless, the existing BA method in IEEE 802.11ad may take as long as several seconds to process a large number of candidate beams [4]. In order to shorten the BA latency, we begin to explore whether there are other methods that can determine the optimal beam without searching the entire beam space.

There are some early efforts on addressing this challenge in the existing literature. Based on the sparse characteristic of the mmWave channel, Marzi et al. developed a compressed sensing BA method [5]. Certain out-of-band information, e.g., the Wi-Fi signal, is exploited to identify the optimal beam in [6]. These works perform BA with the assistance of excessive additional information besides

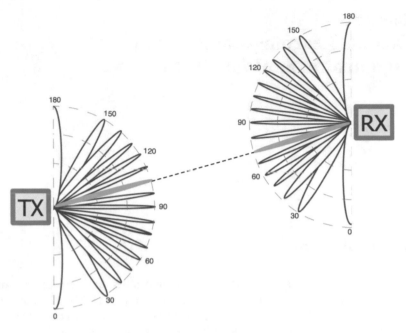

Fig. 3.1 A beam alignment example with 16 beams. The well-aligned transmitter and receiver beams are represented by solid green beams

RSS. Surprisingly, a crucial feature, the correlation structure between beams, is overlooked in previous works. In fact, the RSS of adjacent beams is similar which means adjacent beams are highly correlated. In this way, if a beam performs poorly, its nearby beams are very likely to perform worse. The measurement of one beam reveals not only its own information but also its nearby beams. Therefore, it is possible to identify the optimal beam using the information learned from nearby beams, which avoids searching the entire beam space.

In this chapter, we propose a fast BA algorithm, named hierarchical beam alignment (HBA), by *utilizing the correlation structure among beams and the prior knowledge on the channel fluctuation*. In the BA problem, fast BA means identifying the optimal beam with the minimum latency. This problem boils down to selecting beams sequentially within a certain period to maximize the cumulative RSS, which can be formulated as a stochastic multi-armed bandit (MAB) problem. Our proposed algorithm contains two characteristics that can effectively solve this problem. Firstly, theoretical analysis indicates that the correlation structure among beams in the *multipath* channel can be characterized by a *multimodal* function. Utilizing this correlation structure, the proposed algorithm intelligently narrows the searching space to identify the optimal beam. Secondly, combining the prior knowledge on the channel fluctuation to appropriately accommodate reward uncertainty, the proposed algorithm avoids excessive exploration which further accelerates the BA process. According to the theoretical analysis, the regret of HBA is *bounded*, so the proposed algorithm is asymptotically optimal. Extensive simulation results prove that even if the prior knowledge is coarse, HBA can identify the optimal beam with a high probability and reduce the number of beam measurements in the multipath channel. In particular, compared with the BA method in IEEE 802.11ad, the proposed algorithm reduces the BA latency by several orders of magnitude.

The remainder of this chapter is organized as follows. Section 3.2 reviews the related works. In Sect. 3.3, system model and problem formulation are presented. Section 3.4 proposes a fast BA algorithm. Section 3.5 analyzes the regret performance of the proposed algorithm. Simulation results are given in Sect. 3.6. Finally, Sect. 3.7 summarizes this chapter.

3.2 Related Works on Beam Alignment

The BA problem in mmWave systems has attracted much attention recently. Zhou et al. elaborated on the challenges of the random access protocol in the BA process in dense networks [7]. Besides, the authors developed a solution from the MAC perspective. By taking advantage of the sparse property that only a few paths exist in the mmWave channel, the compressed sensing solution can align beams with low beam measurement complexity of $O(L \log N)$, where L is the number of channel paths and N is the number of beams [5]. The method is suitable for mmWave systems which can obtain the accurate phase information. In another research direction, Wang et al. exploited a fast-discovery multi-resolution beam search in [8], which first detects a wide beam and continues to narrow beams until the best beam is determined. Although feasible, the approach requires adjusting the beam resolution at each step. On the other hand, Xiao et al. proposed a hierarchical codebook search method to efficiently identify the optimal beam through the joint use of sub-array and deactivation technology [9]. In addition, they provide the closed-form representation of the hierarchical codebook. Sun et al. further proposed a low-overhead beam alignment method based on orthogonal pilots for the multiuser mmWave systems [10]. Another solution exploits certain out-of-band information, i.e., the Wi-Fi signal, to identify the optimal beam [6]. Similar works extract spatial information from sub-6 GHz signals to assist BA as well as improve throughput [11, 12]. Recent efforts leverage the multi-armed beam's capability to improve BA performance. Hassanieh et al. proposed a fast BA protocol that scans multiple directions simultaneously [4]. A similar method was developed to transform the problem of identifying the optimal beam to locating the error in linear block codes to reduce BA complexity [13]. The works in [4–13] provide possible solutions for the BA problem in various scenarios. Different from prior works, our work considers the relevant structure among adjacent beams to assist BA process.

Machine learning techniques, especially reinforcement learning and deep learning, have been widely applied in the current advanced wireless networks to address different problems in very recent years [14–16], including network slicing [17] and resource allocation [18, 19]. Among machine learning algorithms, MAB is a low-complexity learning solution. For basic knowledge on MAB theory, the readers are referred to our detailed introduction in Sect. 2.4 in Chap. 2 in this monograph. The MAB theory has been widely applied in wireless networks, such as power allocation in small base stations [20, 21], content placement in edge caching [22, 23], task assignment in mobile crowdsourcing [24], and mobility management in mobile edge computing [25]. Very recently, based on the MAB theory, the BA problem is studied, online decision-making to strike a balance between *exploitation* and *exploration*. Gulati et al. applied the celebrated upper confidence bound (UCB) algorithm in beam selection in traditional MIMO systems [26]. Sim et al. developed an online beam selection algorithm in mmWave vehicular networks based on contextual bandit theory [27]. By learning information from real-time environment, this work improves the throughput of mmWave networks. A pioneering work in [3] exploits a unimodal structure among beams to accelerate the BA process in static environments. This solution focuses on aligning beams in the single-path channel. Another work developed a distributed BA search method based on adversarial bandit theory [28]. These works provide highly relevant insights into the BA problem in mmWave networks via bandit learning theory. Nevertheless, they do not provide a method to align beams quickly and accurately, especially in complex multipath channels. Different from existing works, we focus on the usage of the correlation structure and prior knowledge to accelerate the RSS-only BA process in the multipath channel.

Moreover, to highlight the difference, we compare the proposed BA solution with multiple celebrated existing BA solutions in terms of algorithm complexity, whether only RSS information is required, whether multipath channel is supported, and whether adopting single-beam pattern, as summarized in Table 3.1. The proposed solution has a low complexity while only requiring RSS information and single-beam pattern. In addition, the proposed solution can work in multipath channels.

Table 3.1 A comparison between the proposed BA solution with the existing solutions. "NA" means not available

Work	Complexity	Only RSS	Multipath channel	Single beam
Exhaustive search	$O(N^2)$	Yes	Yes	Yes
802.11ad	$O(N)$	Yes	Yes	Yes
Compressed sensing [5]	$O(L \log N)$	No	Yes	Yes
UBA [3]	$O(N)$	Yes	No	Yes
Out-of-band solutions [12]	NA	No	Yes	NA
Rapid-Link [4]	$O(L \log N)$	Yes	Yes	No
Proposed	$O(\log N)$	Yes	Yes	Yes

3.3 System Model and Problem Formulation

3.3.1 Beam Alignment Model

As shown in Fig. 3.2, we consider a static point-to-point mmWave system in which the transmitter is equipped with N antennas. Uniform linear arrays are assumed to be equipped on both the transmitter and receiver, and each antenna element is connected to a phase shifter to form narrow directional beams [29]. In the BA process, the receiver remains quasi-omni-directional, while the transmitter scans the beam space to identify the best one.

We consider the sparse clustered channel model, i.e., Saleh-Valenzuela model [30]. Suppose that the channel consists of L paths: one dominant LOS path and $L - 1$ NLOS paths, due to strong reflections from the ground or side walls. The channel array response between the transmitter and receiver can be formulated as a mixture of sinusoids

$$h_n = g_0 e^{j \frac{2\pi d}{\lambda} n \vartheta_0} + \sum_{l=1}^{L-1} g_l e^{j \frac{2\pi d}{\lambda} n \vartheta_l} \tag{3.1}$$

where $0 \leq n \leq N - 1$. Let d and λ denote the array element spacing and carrier wavelength, respectively. Typically, $d = \lambda/2$. Let g_0 and g_l represent the channel gains of the LOS path and the l-th NLOS path, respectively. Of particular note is that the channel gain of the LOS path is about 10 dB higher than that of the NLOS path [31]. Let θ denote the physical angle of the channel. The corresponding spatial angle of the channel is denoted by $\vartheta = \cos \theta$. We vectorize the sinusoids $e^{j2\pi dn\vartheta/\lambda}$, $0 \leq n \leq N - 1$ into a vector $\mathbf{x}(\vartheta) \in \mathbb{C}^{N \times 1}$. Thus, the channel vector is given by

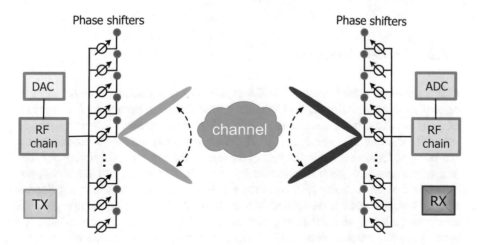

Fig. 3.2 The point-to-point mmWave system

$$\mathbf{h} = g_0\mathbf{x}(\vartheta_0) + \sum_{l=1}^{L-1} g_l\mathbf{x}(\vartheta_l) \in \mathbb{C}^{N\times 1}. \qquad (3.2)$$

Let

$$\mathbf{W} = [\mathbf{w}_1, \mathbf{w}_2, \ldots, \mathbf{w}_N] \in \mathbb{C}^{N\times N}$$

denote the unitary discrete Fourier transform (DFT) matrix whose columns constitute the transmit beam space, given by

$$\mathbf{W} = \frac{1}{\sqrt{N}}[\mathbf{x}(\omega_1), \mathbf{x}(\omega_2), \ldots, \mathbf{x}(\omega_N)]. \qquad (3.3)$$

In (3.3), $\omega_i = \frac{2i-N}{N}$ represents the spatial angle of the i-th beam [5]. According to the BA method in IEEE 802.11ad, the transmitter scans all the beams in $\mathbf{W}$, while the receiver beam keeps omni-directional. The received signal vector is given by

$$\mathbf{y} = \sqrt{P}\mathbf{h}^H\mathbf{W} + \mathbf{n} \qquad (3.4)$$

where $\mathbf{n}$ denotes the additive Gaussian white noise vector. Let N_oW represent the mean noise power, where W is the channel bandwidth and N_o is the noise power density.

The problem of identifying the optimal transmit beam boils down to the problem of identifying the element with the maximum magnitude within $\mathbf{y}$. Therefore, in order to identify the optimal beam, the BA method in IEEE 802.11ad protocol needs to measure the RSS of all the transmit beams, resulting in a high beam measurement complexity [4]. Searching the entire beam space incurs a remarkable BA delay, especially when the beam space is large.

3.3.2   Problem Statement

In this subsection, we formulate the BA problem as a stochastic MAB problem in case of *stationary* environment. Consider a time slotted system with T time slots of equal duration. In time slot $t \in \{1, 2, \ldots, T\}$, the transmitter selects a beam to transmit data. Let $\mathcal{B} = \{b_1, b_2, \ldots, b_N\}$ denote the set of candidate beams, which can be considered as *arms* in the bandit theory. At the beginning of time slot t, the transmitter selects a beam represented by $b^t \in \mathcal{B}$. At the end of time slot t, the transmitter observes noisy RSS from the receiver, i.e., $r(b^t)$, which is considered as a *reward*. Rigorously, the reward is a random variable due to uncertain channel conditions, such as shadow fading and the disturbance effect. For simplicity, we assume that the reward follows a Gaussian distribution with a variance σ^2. In other words, σ^2 also represents the variance of the channel fluctuation, which is utilized as *prior knowledge* in the following algorithm design. Note that the proposed algorithm can also be applied to non-Gaussian distribution settings, as validated in Sect. 3.6.

Let $b^{1:t} = \{b^1, b^2, \ldots, b^t\}$ denote the sequentially selected beams up to time slot t. The set of corresponding sequential rewards is represented by $r^{1:t} = \{r(b^1), r(b^2), \ldots, r(b^t)\}$. In the MAB setup, the sequential beam selection *policy* is how the transmitter selects the next beam based on previously selected beams $b^{1:t}$ and observed rewards $r^{1:t}$. Let Π be the set of all possible sequential beam selection policies. Our goal is to find a policy, $\pi \in \Pi$, that maximizes the expected cumulative reward (RSS) within a given time horizon of T slots, i.e., $\sum_{t=1}^{T} r(b^t)$. This goal is in line with our target because the fast BA algorithm is to identify the optimal beam with the least latency.

In the MAB theory, *expected cumulative regret* is commonly adopted to evaluate the performance of a given strategy, which represents the expected cumulative difference between the reward of the selected beam and the maximum reward achieved by the optimal beam. The *expected cumulative regret* is defined as

$$
R^\pi(T) = \mathbb{E}\left[\sum_{t=1}^{T} \left(r(b^\star) - r(b^t) \right) \right]
$$
$$
= T \cdot \mathbb{E}\left[r\left(b^\star \right) \right] - \sum_{b_i \in \mathcal{B}} N_{b_i}^\pi(T) \mathbb{E}\left[r\left(b_i \right) \right]
$$

(3.5)

where $b^\star$ represents the optimal beam and $N_{b_i}^\pi(T)$ denotes the number of times that b_i has been selected up to time slot T. Hence, maximizing the cumulative reward is equivalent to minimizing the *expected cumulative regret* within T [3], which can be expressed as

$$
\mathcal{P}1 : \min_{\pi \in \Pi} \quad R^\pi(T)
$$

$$
\text{s.t.} \quad \sum_{b_i \in \mathcal{B}} N_{b_i}^\pi(T) \leq T \tag{3.6a}
$$

$$
N_{b_i}^\pi(T) \in \mathbb{Z}, \forall b_i \in \mathcal{B}. \tag{3.6b}
$$

The problem $\mathcal{P}1$ can be solved by the celebrated UCB algorithm [32]. However, there are twofold critical characteristics that the UCB algorithm does not account for. Firstly, since the RSS of nearby beams are highly correlated, the relevant information of the nearby beams can be used to select the next beam efficiently. Secondly, the prior knowledge on the channel fluctuation reflects the environmental information, which can be used to address the uncertainty of rewards, thereby further accelerating the BA process. Next, we use these two features to accelerate the convergence speed, hence reducing the BA latency.

3.4 Fast Beam Alignment Scheme

In this section, we first analyze and verify that the mean reward (i.e., RSS) over the beam space follows a multimodality structure, which characterizes the inherent correlation among beams. Next, we propose a fast BA algorithm to identify the optimal beam by using the correlation structure and the prior knowledge.

3.4.1 Correlation Structure Among Beams

Consider a *cyclic* undirected graph $G = (\mathcal{B}, E)$, and its vertex $\mathcal{B}$ stands for the beams. Let $(b_i, b_{i+1}) \in E$ denote the edge that connects neighboring beams b_i and b_{i+1}. In addition, $(b_N, b_1) \in E$ indicates that the last beam b_N and the first beam b_1 are neighbors since their beam orientations are close to each other. The unimodality structure is defined as follows.

Definition 3.1 (Unimodality) Let b_{i*} denote the optimal beam in G. The *unimodality* structure indicates that, $\forall b_i \in \mathcal{B}$, there exist a path, $(b_i, b_{i+1}, \ldots, b_{i*})$, along which the mean reward is strictly increasing.

In other words, the unimodality structure means that there is no local optimal beam over the beam space. Then, our purpose is to prove that the correlation structure among beams follows the above unimodality structure. Consider the single-path channel, where g and ϑ represent the channel gain and channel spatial angle of the path, respectively. With (3.4), the mean RSS is given by

$$
\begin{aligned}
\mathbb{E}\left[r(b_i)\right] &= P \left|\mathbf{h}^H \mathbf{w}_i\right|^2 + N_o W \\
&= \frac{Pg^2}{N} \left|\mathbf{x}^H(\vartheta)\mathbf{x}(\omega_i)\right|^2 + N_o W \\
&= \frac{Pg^2}{N} \left|\sum_{n=0}^{N-1} e^{j\frac{2\pi d}{\lambda} n(\omega_i - \vartheta)}\right|^2 + N_o W \\
&= \frac{Pg^2}{N} D(\omega_i - \vartheta) + N_o W, \forall b_i \in \mathcal{B}
\end{aligned}
\tag{3.7}
$$

where

$$
D(x) = \frac{\sin^2(N\pi dx/\lambda)}{\sin^2(\pi dx/\lambda)}
\tag{3.8}
$$

denotes the antenna directivity function, which depends on the angular misalignment x. Hence, the mean RSS is a function of angular misalignment $\omega_i - \vartheta$.

Theorem 3.1 *In the single-path channel, the mean reward (RSS) over the beam space is a unimodal function.*

Proof According to (3.7), the maximum RSS can be achieved with the minimum angular misalignment denoted by, $\delta = \omega_{i^\star} - \vartheta$, where $\omega_{i^\star}$ is the spatial angle for the optimal transmit beam. Hence, $D\,(\omega_i - \vartheta)$ can be rewritten as

$$
D\,(\omega_i - \vartheta) = D\left(\delta + \frac{2(i - i^\star)}{N}\right)
$$

$$
= \frac{\sin^2(N\pi d\delta/\lambda)}{\sin^2\left(\pi d\left(\delta + \frac{2(i-i^\star)}{N}\right)/\lambda\right)}, \forall b_i \in \mathcal{B}. \tag{3.9}
$$

From (3.9), $D\,(\omega_i - \vartheta)$ monotonically increases in $[i^\circ, i^\star]$ and decreases in $[i^\star, i^\star + \frac{N}{2}]$, where $i^\circ = i^\star - \frac{N}{2}$. Hence, the mean RSS function over the beam space increases along path $(b_{i^\circ}, b_{i^\circ+1}, \ldots, b_{i^\star})$ and decreases along path $(b_{i^\star}, b_{i^\star+1}, \ldots, b_{i^\circ-1})$, i.e., $r(b_{i^\circ}) < r(b_{i^\circ+1}) < \ldots < r(b_{i^\star}) > \ldots > r(b_{i^\circ-2}) > r(b_{i^\circ-1})$. With the definition of the unimodality structure, the mean RSS function is unimodal over the beam space in the single-path channel. Hence, Theorem 3.1 is proved. □

The linear combination of several unimodal functions is a *multimodal* function, which means that there exist several local optimums.

Corollary 3.1 *In the multipath channel, the mean reward (RSS) over the beam space is a multimodal function. The dominant peak of the multimodal function is caused by the LOS path, while other peaks are caused by NLOS paths.*

Proof Similar to (3.7), the mean of the RSS in the multipath channel is represented by

$$
\mathbb{E}\,[r(b_i)] = \underbrace{\frac{Pg_0^2}{N}D\,(\omega_i - \vartheta_0)}_{\text{LOS component}} + \underbrace{\sum_{l=1}^{L-1}\frac{Pg_l^2}{N}D\,(\omega_i - \vartheta_l)}_{\text{NLOS component}} + N_o W \tag{3.10}
$$

From this equation, it can be observed that the aggregated RSS consists of one LOS component and several NLOS components. For each individual path of the mmWave channel, the corresponding RSS function is a unimodal function based on Theorem 3.1. Therefore, the RSS function in the multipath channel is a collection of several unimodal functions, which can be considered as a multimodal function. In particular, there are L paths in the mmWave channel, corresponding to L peaks in the multimodal function. As the channel gain of the LOS path is significantly larger than that of NLOS paths, i.e., $g_0^2 > g_l^2$. Hence, the dominant peak corresponds to the LOS path, while other peaks correspond to NLOS paths. Thus, Corollary 3.1 is proved. □

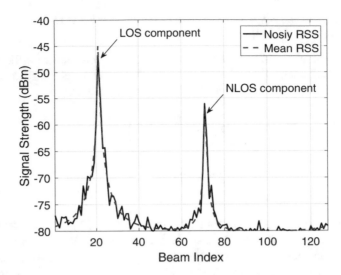

Fig. 3.3 The RSS function over the beam space in a two-path channel with 128 beams. The peak value generated by the LOS link is around 10 dB higher than that generated by the NLOS link

Figure 3.3 illustrates the RSS function over the beam space in a two-path channel. Actually, the RSS is noisy because of the channel fluctuation. We observe that the mean RSS function follows the multimodality structure. For a two-path mmWave channel, there are two peaks in the mean RSS function: the dominant peak is due to the LOS path, and another smaller peak is due to the NLOS path. Moreover, the multimodality structure has been observed in many in-field measurements in mmWave systems, which further validates our theoretical results.

Remark 3.1 According to theoretical analysis, we can conclude that the RSS depends on the angular misalignment. The angular misalignment varies gradually across adjacent beams, so the values of the RSS of nearby beams are close to each other; as a result, neighboring beams are highly correlated. Besides, due to the multipath nature of channel, the RSS function exhibits the multimodality structure, which can be utilized to accelerate the convergence speed of the BA process.

3.4.2 Prior Knowledge of mmWave Networks

In addition to the aforementioned correlation structure, other prior knowledge can also be exploited to further accelerate the BA process. Since the reward is affected by wireless environments, the channel fluctuation statistics reflects the underlying information of the wireless environments. The utilization of channel fluctuation statistics can help better accommodate the reward uncertainty, thereby reducing the exploration. Specifically, we assume that the channel fluctuation variance σ^2

is known *a priori* to speed up the BA process. In practice, such prior knowledge can be obtained in the system initialization phase before the BA process is invoked. Practical mmWave systems also collect the variance of channel fluctuation periodically. Besides, since the channel statistical information changes slowly in static environments, there is no need to frequently collect the information. Note that the proposed algorithm works even with coarse prior knowledge at the expense of slower convergence or lower beam detection accuracy, which is presented in Sect. 3.6.

3.4.3 Learning-Based Beam Alignment Algorithm

As mentioned earlier, the mean reward function exhibits the multimodality structure, so we improve and extend the hierarchical optimistic optimization (HOO) algorithm [33] to the BA problem. Due to the lack of prior knowledge, HOO develops a large confidence margin to adopt to the reward uncertainty, which leads to slower convergence. Similar to the famous Bayesian principles [34], we use the prior knowledge to obtain an appropriate confidence margin to further accelerate the convergence without unnecessary exploration. The proposed HBA algorithm is sketched in Algorithm 1. Specifically, $Ber(p)$ represents a Bernoulli distribution with a parameter p, and $leaf(\mathcal{T})$ represents the leaf node of a tree $\mathcal{T}$ in the algorithm.

Theoretical analysis shows that if a beam performs well, its nearby beams are also highly likely to perform well. The proposed algorithm is built upon the correlation structure among beams, and its core idea is to conduct dense exploration around good beams and loose exploration in others. To this end, a search tree is constructed, the nodes of which are associated with search regions. Deeper nodes represent smaller search regions, as an illustrative example shown in Fig. 3.4. The algorithm runs in discrete time slots and constructs a binary tree incrementally. At each time slot, the node selection process selects a new node and adds it to the search tree. Once selected, the beam located in the selected node is measured, and then the corresponding reward is observed. Afterward, the attributes of the search tree are updated based on the newly observed reward. In this way, the algorithm intelligently narrows the search region until the optimal beam is identified. It is worth noting that selecting a new node means exploring the region related to the node, while the search tree explores the region based on previously selected beams and observed rewards.

Next, we elaborate on the algorithm. In the initialization phase, the beam space, $\mathcal{B}$, is mapped to a region $\mathcal{X} = [0, 1]$, which is evenly partitioned by each beam. Similarly, the RSS function, $r(b_i), \forall b_i \in \mathcal{B}$, is mapped to a normalized reward function, $f(x), \forall x \in \mathcal{X}$, within $[0, 1]$. At the beginning, the search tree $\mathcal{T}$ only contains a root node $(0, 1)$. The node in the tree is represented by (h, j), where h denotes the depth from the root node and $j, 1 \leq j \leq 2^h$ denotes the index at depth h. In addition, each node in the tree is associated with a region. Let $C_{h,j}$

Algorithm 1 HBA algorithm

Input: ζ, ρ_1, γ, and σ^2;
Output: $b^\star$;
 1: Initialization: Set $\mathcal{T} = \{(0, 1)\}$, $Q_{2,1} = Q_{2,2} = +\infty$, $x_L = 0$, and $x_H = 1$;
 2: $(h, j) \leftarrow (0, 1)$, $\mathcal{P} \leftarrow \{(h, j)\}$;
 3: ▷ New node selection
 4: **while** $(h, i) \in \mathcal{T}_t$ **do**
 5:         **if** $Q_{h+1,2j-1}(t) > Q_{h+1,2j}(t)$ **then**
 6:                 $(h, j) \leftarrow (h + 1, 2j - 1)$;
 7:                 Update $x_L = x_a$;
 8:         **else if** $Q_{h+1,2j-1}(t) < Q_{h+1,2j}(t)$ **then**
 9:                 $(h, j) \leftarrow (h + 1, 2j)$;
10:                 Update $x_H = x_a$;
11:         **else**
12:                 $(h, j) \leftarrow (h + 1, 2j - Ber(0.5))$;
13:                 Update the search region;
14:         **end**
15:         $(H_t, J_t) \leftarrow (h, j)$;
16:         $\mathcal{T}_{t+1} = \mathcal{T}_t \cup \{(H_t, J_t)\}$;
17: **end**
18: $(H_t, J_t) \leftarrow (h, j)$;
19: $\mathcal{T}_{t+1} = \mathcal{T}_t \cup \{(H_t, J_t)\}$;
20: ▷ Attributes update
21: Measure the beam located in the center C_{H_t,J_t} and observe the reward r^t;
22: Update $N_{h,j}(t)$ and $R_{h,j}(t)$, $\forall(h, j) \in \mathcal{P}$, with (3.11) and (3.12), respectively;
23: Update $E_{h,j}(t)$, $\forall(h, j) \in \mathcal{T}_t$, with (3.13);
24: $Q_{H+1,2J-1}(t) = Q_{H+1,2J}(t) = +\infty$;
25: $\hat{\mathcal{T}} = \mathcal{T}_t$;
26: **for** $(h, j) \in \hat{\mathcal{T}}$ **do**
27:         $(h, j) \leftarrow leaf(\hat{\mathcal{T}})$;
28:         Update $Q_{h,j}(t)$ with (3.14);
29:         $\hat{\mathcal{T}} \leftarrow \hat{\mathcal{T}} \setminus (h, j)$;
30: **end**
31: ▷ Terminating condition
32: **if** $x_H - x_L < \zeta/N$ **then**
33:         Terminate beam search and select current beam $b^\star$;
34: **end**

---

represent the region of (h, j). Specifically, the root node represents the entire region, i.e., $C_{0,1} = [0, 1]$. Let $(h + 1, 2j - 1)$ and $(h + 1, 2j)$ denote the left and the right child node of (h, j), respectively. Two child nodes partition the region of their parent node. Consider $C_{h,j} = [x_L, x_H]$: the left child node is associated with a region $C_{h+1,2j-1} = [x_L, x_a]$, and the right child node is associated with a region $C_{h+1,2j} = [x_a, x_H]$, where $x_a = x_L + (x_H - x_L)/2$ is the middle point of $C_{h,j}$. At time slot t, HBA consists of the following three phases:

1. *New node selection.* In this phase, a new node is firstly selected. Let $\mathcal{T}_t$ denote the tree at time t. Starting from the root node, we compare the estimated rewards (denote by Q-values) of two children at each time slot until a new node $(H_t, J_t) \notin$

$\mathcal{T}_t$ is selected. Specifically, HBA traverses the tree first and selects the child with a higher Q-value and breaks ties randomly (Lines 4–14). The selected node is added to the tree, i.e., $\mathcal{T}_{t+1} = \mathcal{T}_t \cup \{(H_t, J_t)\}$, and the path from the root node to the selected node is stored in $\mathcal{P}$.

2. *Attributes update.* This stage is performed to update the attributes of all nodes in the tree. For the selected node in the previous phase, measure the beam at the center of C_{H_t, J_t}, and get the corresponding reward r_t. According to the newly observed reward, the estimated mean reward $Q_{h,j}(t)$ is updated as follows.

- Firstly, as the new node is the descendant of all the nodes in path $\mathcal{P}$, $N_{h,j}(t)$, which represents the number of times that (h, j) has been selected until time slot t, is updated by

$$N_{h,j}(t) = N_{h,j}(t-1) + 1, \forall(h, j) \in \mathcal{P}. \tag{3.11}$$

- Secondly, $R_{h,j}(t)$ represents the mean measured reward of (h, j) up to time slot t, which is updated by

$$R_{h,j}(t) = \frac{\left(N_{h,j}(t) - 1\right) R_{h,j}(t-1) + r^t}{N_{h,j}(t)}, \forall(h, j) \in \mathcal{P}. \tag{3.12}$$

- Thirdly, for each node in the tree, the initial estimated reward $E_{h,j}(t)$ is updated by

$$E_{h,j}(t) = \begin{cases} R_{h,j}(t) + \sqrt{\frac{2\sigma^2 \log t}{N_{h,j}(t)}} + \rho_1 \gamma^h, & \text{if } N_{h,j}(t) > 0 \\ +\infty, & \text{otherwise} \end{cases} \tag{3.13}$$

where $\sqrt{\frac{2\sigma^2 \log t}{N_{h,j}(t)}}$ is the confidence margin for adapting the uncertainty of rewards. As mentioned before, we use the prior knowledge of the variance of channel fluctuation and adopt the Bayesian principle to design the confidence margin. The term $\rho_1 \gamma^h$ accounts for the maximum variation of the mean reward function over $C_{h,j}$, where $\rho_1 > 0$ and $\gamma \in (0, 1)$. This term is due to the bounded diameter assumption, which is discussed later in Sect. 3.5. The values of ρ_1 and γ are selected based on extensive simulation trials. For a binary tree case, the value γ is typically set to be 0.5 [33]. It should be noted that e-values of all the unexplored nodes are set to infinity.

- Finally, for each node in the tree, the estimated mean reward, $Q_{h,j}(t)$, should be recursively updated through the following bound:

$$Q_{h,j}(t) = \begin{cases} \min\{E_{h,j}(t), \max\{Q_{h+1,2j-1}(t), Q_{h+1,2j}(t)\}\}, & \text{if } N_{h,j}(t) > 0 \\ +\infty, & \text{otherwise.} \end{cases} \tag{3.14}$$

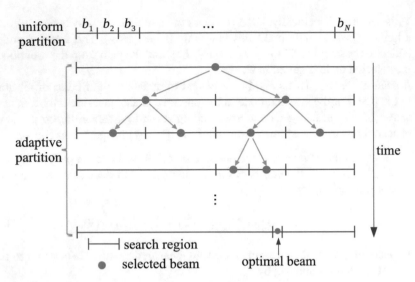

Fig. 3.4 The proposed algorithm operates in a "zooming" manner

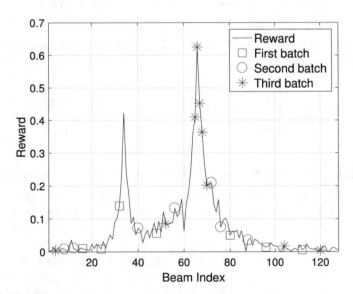

Fig. 3.5 The region that contains the dominant peak is explored intensively, while others are explored loosely

This bound depends on two terms. The first term, $E_{h,j}(t)$, is an upper bound for $Q_{h,j}(t)$ due to the definition of E-values. The second term, i.e.,

$$\max\{Q_{h+1,2j-1}(t), Q_{h+1,2j}(t)\}$$

is another valid upper bound of $Q_{h,j}(t)$. Since $C_{h,j} = C_{h+1,2j-1} \cup C_{h+1,2j-1}$, the maximum value between the Q-values in two subsets is the upper bound of Q-value in the union set. Combining these two items can get a tighter upper bound by taking the minimum of these two items. Note that Q-values should be updated from the leaf node of the tree, because Q-values of child nodes form the upper bound of their parent node (Lines 21–27).

3. *Terminating condition*. Over time, the depth of the tree increases and the search area gradually shrinks. When the search region is sufficiently small, i.e., $x_H - x_L < \zeta/N$ where $0 < \zeta < 1$, the BA process is terminated, and the beam located in the final region is selected as the optimal beam. The value of ζ should be carefully selected on the basis of extensive simulation trials. It is worth noting that the larger the value of ζ, the faster the convergence speed and the lower the beam detection accuracy.

Remark 3.2 For better understanding of the HBA, we provide two illustrative examples as follows:

1. As shown in Fig. 3.4, HBA operates similarly to a "zooming process." At the beginning, the search region is the entire region evenly divided by the beams. As time goes by, the search region is adaptively partitioned, and the algorithm gradually zooms to the region containing the optimal beam.
2. The sequentially selected beams in the BA process are depicted in Fig. 3.5. The selected beams are divided into three batches according to the timeline. The first batch of beams are randomly positioned throughout the area. The second batch of beams gets closer to the dominant peak. The last batch is mainly concentrated around the optimal beam. We notice that the algorithm performed intensive exploration in areas containing good beams, while it performed less exploration in other areas.

3.5 Theoretical Analysis

3.5.1 Algorithm Complexity Analysis

At time slot T, $\mathcal{T}_t$ contains T nodes as the tree increments by one node at each time slot. Therefore, the storage complexity of the proposed algorithm is linear, i.e., $O(T)$. In addition, the attributes of all the nodes in the tree should be updated at each time slot, and hence the running time at each time slot is also linear. As the algorithm runs T time slots, the computational complexity of the HBA algorithm is quadratic, i.e., $O(T^2)$. Because of the terminating condition, the tree is a finite tree, so the storage complexity and computational complexity are bounded.

3.5.2 Cumulative Regret Performance Analysis

In this section, we analyze the upper bound on the cumulative regret for the proposed algorithm. In order to facilitate regret analysis, we make the following two assumptions.

Assumption Weak Lipschitz: For any x around the optimal $x^\star$, there exist constants $c_H > 0$ and $\alpha > 0$ such that

$$f^\star - f(x) \le c_H \|x^\star - x\|^\alpha. \tag{3.15}$$

This hypothesis shows that the reward function satisfies the week Lipschitz condition, which can avoid the sharp valleys of high regret near the optimal point. Moreover, the weak Lipschitz condition is mild, which only affects the region near the optimal value. This assumption has been proved in many practical applications [21].

Assumption

1. **Bounded diameter**: For a region $C_{h,j}$, of depth h, the diameter of the region is defined as $D(C_{h,j}) = \max\limits_{x,y \in C_{h,j}} q(x,y)$, where $q(x,y) = w\|x-y\|^\beta$ represents the dissimilarity between x and y. The diameter of the region is upper bounded by $\rho_1 \gamma^h$ for constants $\rho_1 > 0$ and $0 < \gamma < 1$.
2. **Well-shaped region**: For a region $C_{h,j}$, of depth h, the region contains a ball with a radius of $\rho_2 \gamma^h$ which locates in the center of $C_{h,j}$.

The bounded diameter condition is to upper bound the maximum variation of $f(x)$ within the region $C_{h,j}$. On the contrary, the well-shaped region condition is to lower bound the minimum variation of $f(x)$ within the region $C_{h,j}$. Note that any region in the reward function satisfies the bounded diameter and well-shaped region conditions [33]. These conditions are used to derive the accumulated regret in the following analysis.

Definition 3.2 ϵ-**optimal**: Let $f^\star_{h,j} = \max\limits_{x \in C_{h,j}} f(x)$ be the optimal reward in $C_{h,j}$. If $f^\star_{h,j} > f^\star - \epsilon_{h,j}$, $C_{h,j}$ is the $\epsilon_{h,j}$-optimal region.

For example, if $\epsilon_{h,j} = 0$, $C_{h,j}$ is the optimal region where the optimal value $x^\star$ locates. Otherwise, if $\epsilon_{h,j} > 0$, $C_{h,j}$ is a suboptimal region. Let $\epsilon_{h,j}$ represent the *suboptimality* of (h, j).

To obtain the regret bound, we first provide the following lemma.

Lemma 3.1 *For any node* (h, j) *whose suboptimality is larger than* $\rho_1 \gamma^h$, *the expected number of times that* (h, j) *has been visited until time slot* T *is upper bounded by*

$$\mathbb{E}\left[N_{h,j}(T)\right] \leq \frac{8\sigma^2 \log T}{\left(\epsilon_{h,j} - \rho_1 \gamma^h\right)^2} + c \tag{3.16}$$

where c *is a constant.*

Proof For any integer $m > 0$, according to the definition, the average times that node (h, j) has been visited up to time slot T is given by

$$
\begin{aligned}
\mathbb{E}\left[N_{h,j}(T)\right] &= \mathbb{E}\left[\sum_{t=1}^{T} \mathbb{1}_{(H_t, J_t) \in C_{h,j}}\right] \\
&= \mathbb{E}\left[\sum_{t=1}^{T} \mathbb{1}_{\{(H_t, J_t) \in C_{h,j}, N_{h,j}(t) \leq m\}}\right] \\
&\quad + \mathbb{E}\left[\sum_{t=1}^{T} \mathbb{1}_{\{(H_t, J_t) \in C_{h,j}, N_{h,j}(t) > m\}}\right] \\
&\leq m + \mathbb{E}\left[\sum_{t=m+1}^{T} \mathbb{1}_{\{(H_t, J_t) \in C_{h,j}, N_{h,j}(t) > m\}}\right] \\
&= m + \sum_{t=m+1}^{T} \mathbb{P}\left((H_t, J_t) \in C_{h,j}, N_{h,j}(t) > m\right).
\end{aligned}
\tag{3.17}
$$

where $\mathbb{1}_{\{\cdot\}}$ is the indicator function and $(H_t, J_t) \in C_{h,j}$ denotes the selected node (H_t, J_t) located within $C_{h,j}$. The first equality holds because $N_{h,j}(t) > m$ only occurs when t is larger than m.

We apply a case study to obtain an upper bound of $\mathbb{E}\left[N_{h,j}(T)\right]$. Assume node (h, j) is selected at time slot t. The path from root node $(0, 1)$ to (h, j) is given by

$$\mathcal{P} = \{(0, 1), (1, j_1^{\star}), \ldots, (k, j_k^{\star}), (k + 1, j_{k+1}^{o}), \ldots, (h, j)\}$$

where k denotes the largest depth of the optimal node in the path. Before node $(k, j_k^{\star})$, the optimal nodes are selected. For notation simplicity, we omit the time slot t in $Q_{k,j}(t)$. After traversing node $(k, j_k^{\star})$, a suboptimal node $(k + 1, j_{k+1}^{o})$ is selected instead of the optimal node $(k + 1, j_{k+1}^{\star})$, because the suboptimal node has a larger Q-value than the optimal node, i.e., $Q_{k+1,j^o} \geq Q_{k+1,j^{\star}}$. As Q-values increase along path $\mathcal{P}$, we have

$$Q_{k+1,j^{\star}} \leq Q_{k+1,j_{k+1}^{o}} \leq, \ldots, \leq Q_{h,j}.$$

Note that Q-values are upper bounded by E-values according to the definition, such that $Q_{k+1,j^\star} \leq E_{h,j}$. Further, event $Q_{k+1,j^\star} \leq E_{h,j}$ can be interpreted as the union of two events, $\{Q_{k+1,j^\star} \leq f^\star\} \cup \{E_{h,j} \geq f^\star\}$. Hence, the probability that (H_t, J_t) locates within $C_{h,j}$ is upper bounded by

$$\mathbb{P}\left((H_t, J_t) \in C_{h,j}\right) \leq \mathbb{P}\left(Q_{k+1,j^\star} \leq f^\star\right) + \mathbb{P}\left(E_{h,j} \geq f^\star\right). \tag{3.18}$$

With the definition of Q-value, the Q-value of a node is the minimum value among the E-value of the node and Q-values of its child nodes. Hence, event $\{Q_{k+1,j^\star} \leq f^\star\}$ can be interpreted as the union of two new events, i.e.,

$$\{E_{k+1,j^\star} \leq f^\star\} \cup \{Q_{k+2,j_{k+2}^\star} \leq f^\star\}.$$

Event $\{Q_{k+2,j_{k+2}^\star} \leq f^\star\}$ can be further recursively expanded as

$$\bigcup_{s=k+2}^{t-1} \{E_{s,j_s^\star} \leq f^\star\}.$$

Hence, we have

$$\mathbb{P}\left(Q_{k+1,j^\star} \leq f^\star\right) \leq \sum_{s=k+1}^{t-1} \mathbb{P}\left(E_{s,j_s^\star} \leq f^\star\right). \tag{3.19}$$

Substituting (3.19) and (3.18) into (3.17), we have

$$\mathbb{E}\left[N_{h,j}(T)\right] \leq m + \sum_{t=m+1}^{T} \left(\sum_{s=k+1}^{t-1} \mathbb{P}\left(E_{s,j^\star}(t) \leq f^\star\right)\right.$$
$$\left. + \mathbb{P}\left(E_{h,j}(t) \geq f^\star, N_{h,j}(t) > m\right)\right). \tag{3.20}$$

The following analysis is to bound the three terms in (3.20) separately.

Firstly, since m is an arbitrary integer, taking m as the smallest integer that satisfies the condition

$$m \geq \frac{8\sigma^2 \log T}{\left(\epsilon_{h,j} - c_1 \gamma^h\right)^2}.$$

Hence, m is bounded by

$$m \leq \frac{8\sigma^2 \log T}{\left(\epsilon_{h,j} - \rho_1 \gamma^h\right)^2} + 1. \tag{3.21}$$

Secondly, we aim to bound the first term $\mathbb{P}\left(E_{s,j^*} \leq f^*\right)$. For the optimal nodes (h, j^*), according to the definition of E-values, $E_{h,j^*} = \infty$ when $N_{h,j^*} = 0$. Hence, event $E_{h,j^*} \leq f^*$ only occurs when $N_{h,j} \geq 1$. As a result, $\mathbb{P}\left(E_{h,j^*} \leq f^*\right)$ can be rewritten as

$$
\mathbb{P}\left(E_{h,j^*} \leq f^*, N_{h,j} \geq 1\right)
$$

$$
= \mathbb{P}\left(R_{h,j^*} + \sqrt{\frac{2\sigma^2 \log t}{N_{h,j^*}}} + \rho_1 \gamma^h \leq f^*, N_{h,j^*} \geq 1\right)
$$

$$
= \mathbb{P}\left(\left(f^* - R_{h,j^*} - \rho_1 \gamma^h\right) N_{h,j^*} \geq \sqrt{2\sigma^2 N_{h,j^*} \log t}, N_{h,j^*} \geq 1\right)
$$

$$
\overset{(a)}{=} \mathbb{P}\left(\sum_{s=1}^{t}\left(f^* - f(X_s) + \rho_1 \gamma^h\right) \mathbb{1}_{(H_t, J_t) \in C_{h,j^*}}\right.
$$

$$
\left. + \sum_{s=1}^{t}(f(X_s) - Y_s) \mathbb{1}_{(H_t, J_t) \in C_{h,j^*}} \geq \sqrt{2\sigma^2 N_{h,j^*} \log t}, N_{h,j^*} \geq 1\right)
$$

$$
\overset{(b)}{\leq} \mathbb{P}\left(\sum_{s=1}^{t}(f(X_s) - Y_s) \mathbb{1}_{(H_t, J_t) \in C_{h,j^*}} \geq \sqrt{2\sigma^2 N_{h,j^*} \log t}, N_{h,j^*} \geq 1\right)
$$

$$
\overset{(c)}{=} \mathbb{P}\left(\sum_{p=1}^{N_{h,j^*}} \left(\tilde{Y}_p - f(\tilde{X}_p)\right) \geq \sqrt{2\sigma^2 N_{h,j^*} \log t}, N_{h,j^*} \geq 1\right).
$$

(3.22)

In (3.22), the first step follows from the definition of E-value in (3.13); (a) is obtained from the definition of N_{h,j^*}, where $X_s, \forall s = 1, 2, \ldots, t-1$ denotes the sequentially selected beams up to time $t-1$ and the corresponding reward sequence is represented by Y_s; (b) follows from the fact that $f^* - f(X_t) - \rho_1 \gamma^h < 0$ holds for all the beams in the optimal region C_{h,j^*}; (c) is because the definition of a new beam selection sequence $\hat{X}_p, \forall p = 1, 2, 3, \ldots$ whose corresponding reward sequence is $\hat{Y}_p$.

Let $T_p = \min\{t : N_{h,j}(t) = p\}$ represent the time sequence for the selected node in $C_{h,j}$. The sequentially selected beams can be represented by a new sequence $\hat{X}_p = X_{T_p}, \forall p = 1, 2, 3, \ldots$, and (3.22) can be further bounded by

$$\mathbb{P}\left(\sum_{p=1}^{N_{h,j_h^\star}} \left(\tilde{Y}_p - f(\tilde{X}_p)\right) \geq \sqrt{2\sigma^2 N_{h,j^\star} \log t}, N_{h,j_h^\star} \geq 1\right)$$

$$\overset{(a)}{\leq} \sum_{s=1}^{t} \mathbb{P}\left(\sum_{p=1}^{s} \left(\tilde{Y}_p - f(\tilde{X}_p)\right) \geq \sqrt{2\sigma^2 s \log t}\right) \tag{3.23}$$

$$\overset{(b)}{\leq} \sum_{s=1}^{t} \exp\left(-\frac{4\sigma^2 s \log t}{s\sigma^2}\right) = t^{-3}.$$

In (3.23), (a) can be acquired via the union bound that takes all possible values of $N_{h,j_h^\star}$; as $\tilde{D}_p = \tilde{Y}_p - f(\tilde{X}_p)$ can be considered as martingale differences, (b) is obtained via the Hoeffding-Azuma inequality [33]

$$\mathbb{P}\left(\sum_{p=1}^{k} \tilde{D}_p \geq t\right) \leq \exp\left(-\frac{2t^2}{\sum_{p=1}^{k} \sigma^2}\right). \tag{3.24}$$

Thirdly, for suboptimal nodes (h, j), the upper bound of $\mathbb{P}(E_{h,j} \geq f^\star, N_{h,j} > m)$ can be obtained via a similar method of bounding $\mathbb{P}(E_{h,j^\star} \leq f^\star, N_{h,j} \geq 1)$, such that

$$\mathbb{P}\left(E_{h,j} \geq f^\star, N_{h,j} > m\right)$$

$$= \mathbb{P}\left(R_{h,j} + \sqrt{\frac{2\sigma^2 \log t}{N_{h,j}}} + \rho_1 \gamma^h \geq f_{h,j}^\star + \epsilon_{h,j}, N_{h,j} > m\right)$$

$$\overset{(a)}{\leq} \mathbb{P}\left(R_{h,j} \geq f_{h,j}^\star + \frac{\epsilon_{h,j} - \rho_1 \gamma^h}{2}, N_{h,j} > m\right)$$

$$= \mathbb{P}\left(\left(R_{h,j} - f_{h,j}^\star\right) N_{h,j} \geq \frac{\epsilon_{h,j} - \rho_1 \gamma^h}{2} N_{h,j}, N_{h,j} > m\right)$$

$$= \mathbb{P}\left(\sum_{s=1}^{t} \left(Y_s - f_{h,j}^\star\right) \mathbb{1}_{(H_s, J_s) \in C_{h,j}} \geq N_{h,j} \frac{\epsilon_{h,j} - \rho_1 \gamma^h}{2}, N_{h,j} > m\right) \tag{3.25}$$

$$\leq \mathbb{P}\left(\sum_{s=1}^{t} (Y_s - f(X_s)) \mathbb{1}_{(H_s, J_s) \in C_{h,j}} \geq N_{h,j} \frac{\epsilon_{h,j} - \rho_1 \gamma^h}{2}, N_{h,j} > m\right)$$

$$\overset{(b)}{=} \mathbb{P}\left(\sum_{p=1}^{N_{h,j}} \left(\hat{Y}_p - f(\hat{X}_p)\right) \geq N_{h,j} \frac{\epsilon_{h,j} - \rho_1 \gamma^h}{2}, N_{h,j} > m\right)$$

In (3.25), (a) due to the substitution of $N_{h,j}(t) \geq \frac{8\sigma^2 \log t}{(\epsilon_{h,j} - \rho_1 \gamma^h)^2}$ where $m \geq$ $\frac{8\sigma^2 \log t}{(\epsilon_{h,j} - \rho_1 \gamma^h)^2}$; (b) is obtained via a similar method as (3.22)(c), where a new beam sequence $\{\hat{X}_1, \hat{X}_2, \ldots, \hat{X}_p\}$ is formed to represent the sequentially selected beams in $C_{h,j}$. Next, (3.25) can be further bounded by

$$
\mathbb{P}\left(\sum_{p=1}^{N_{h,j}} \left(\hat{Y}_p - f(\hat{X}_p)\right) \geq N_{h,j} \frac{\epsilon_{h,j} - \rho_1 \gamma^h}{2}, N_{h,j} > m\right)
$$

$$
\overset{(a)}{\leq} \sum_{k=m+1}^{t} \mathbb{P}\left(\sum_{p=1}^{k} \left(\hat{Y}_p - f(\hat{X}_p)\right) \geq \frac{k(\epsilon_{h,j} - \rho_1 \gamma^h)}{2}\right)
$$

$$
\overset{(b)}{\leq} \sum_{k=m+1}^{t} \exp\left(-\frac{k\left(\epsilon_{h,j} - \rho_1 \gamma^h\right)^2}{2\sigma^2}\right) \tag{3.26}
$$

$$
\leq t \exp\left(-\frac{m\left(\epsilon_{h,j} - \rho_1 \gamma^h\right)^2}{2\sigma^2}\right)
$$

$$
\overset{(c)}{\leq} t \exp\left(-4 \log T\right)
$$

$$
= tT^{-4}.
$$

In (3.26), (a) holds due to a similar union bound in (3.23)(a); (b) is obtained via the Hoeffding-Azuma inequality; (c) is obtained via the substitution of $m \geq \frac{8\sigma^2 \log T}{(\epsilon_{h,j} - \rho_1 \gamma^h)^2}$.

Finally, substituting (3.21), (3.23), and (3.26) into (3.20), the upper bound is given by

$$
\mathbb{E}\left[N_{h,j}(T)\right] \leq \frac{8\sigma^2 \log T}{\left(\epsilon_{h,j} - \rho_1 \gamma^h\right)^2} + 1 + \sum_{t=m+1}^{T} \left(\sum_{k+1}^{t-1} t^{-3} + tT^{-4}\right)
$$

$$
\leq \frac{8\sigma^2 \log T}{\left(\epsilon_{h,j} - \rho_1 \gamma^h\right)^2} + 1 + \sum_{t=1}^{T} \left(t^{-2} + T^{-3}\right) \tag{3.27}
$$

$$
\leq \frac{8\sigma^2 \log T}{\left(\epsilon_{h,j} - \rho_1 \gamma^h\right)^2} + c
$$

where c is a constant. The last step is because $\sum_{t=1}^{T} t^{-2}$ is bounded. Hence, Lemma 3.1 is proved. $\qquad\square$

Remark 3.3 From Lemma 3.1, the number of visits of suboptimal nodes increases logarithmically with time, which means that the cumulative regret of the proposed algorithm is sublinear. In addition, the number of times that a suboptimal node has been visited depends on the variance of the channel fluctuation. A larger variance of the channel fluctuation implies a more noisy wireless environment, resulting in more exploration efforts to eliminate the reward uncertainty.

Based on the above lemma, an upper bound is obtained below.

Theorem 3.2 *The upper bound on the cumulative regret of HBA is*

$$R^{\pi}(T) = O\left(\sqrt{T \log T}\right). \tag{3.28}$$

Proof All nodes with depth h can be divided into two subsets: Φ_h that denotes the set of all the $2\rho_1\gamma^h$-optimal nodes and Ω_h that denotes the set of nodes whose parents belong to Φ_{h-1} while itself does not belong to Φ_h. Let $H \geq 1$ be an integer whose value is determined later. Based on the above definition, $\mathcal{T}$ can be divided into three subtrees: $\mathcal{T}_1$, $\mathcal{T}_2$, and $\mathcal{T}_3$. Let $\mathcal{T}_1$ contain Φ_H and its decedents. Let $\mathcal{T}_2$ include all the $2\rho_1\gamma^h$-optimal nodes at all the depths smaller than H, i.e., $\mathcal{T}_2 = \bigcup_{h=1}^{H-1} \Phi_h$. Let $\mathcal{T}_3$ include all the nodes in Ω_h at all the depths smaller than H, i.e., $\mathcal{T}_3 = \bigcup_{h=1}^{H} \Omega_h$. Hence the cumulative regret can be partitioned as

$$R^{\pi}(T) = \mathbb{E}\left[R^{\pi}(\mathcal{T}_1)\right] + \mathbb{E}\left[R^{\pi}(\mathcal{T}_2)\right] + \mathbb{E}\left[R^{\pi}(\mathcal{T}_3)\right] \tag{3.29}$$

where

$$\mathbb{E}\left[R^{\pi}(\mathcal{T}_i)\right] = \mathbb{E}\left[\sum_{t=1}^{T} \left(f^{\star} - f(X_t)\right) \mathbb{1}_{\{(H_t, J_t) \in \mathcal{T}_i\}}\right].$$

Next, the regret analysis follows the idea of bounding the regret on each subtree separately.

Step 1: Bounding the Regret on $\mathcal{T}_1$ As each node in Φ_H is $2\rho_1\gamma^H$-optimal, all the beams located in Φ_H are $4\rho_1\gamma^H$-optimal, i.e., $f^{\star} - f(X_t) \leq 4\rho_1\gamma^H$, $X_t \in \Phi_H$. In addition, it is obvious that the number of nodes in subtree $\mathcal{T}_1$ is smaller than the time horizon, i.e., $|\mathcal{T}_1| \leq T$ where $|\cdot|$ represents the cardinality operator. Therefore, the regret on $\mathcal{T}_1$ is upper bounded by

$$\mathbb{E}\left[R^{\pi}(\mathcal{T}_1)\right] \leq 4\rho_1\gamma^H T. \tag{3.30}$$

Step 2: Bounding the Regret on $\mathcal{T}_2$ As $\mathcal{T}_2 = \bigcup\limits_{h=1}^{H-1} \Phi_h$ and each beam in Φ_h is $4\rho_1\gamma^h$-optimal, the regret on $\mathcal{T}_2$ can be written as

$$\mathbb{E}\left[R^\pi\left(\mathcal{T}_2\right)\right] \le \sum_{h=1}^{H-1} 4\rho_1\gamma^h|\Phi_h|.$$

Based on the results in [33], we have $|\Phi_h| \le c_1\left(\rho_2\gamma^h\right)^{-\kappa}$ where $\kappa = \frac{1}{\beta} - \frac{1}{\alpha}$. Specifically, α and β are given in the weak Lipschitz assumption and bounded diameter assumption, respectively. The regret on $\mathcal{T}_2$ can be further bounded by

$$\mathbb{E}\left[R^\pi\left(\mathcal{T}_2\right)\right] \le \sum_{h=1}^{H-1} 4\rho_1\gamma^h c_1\left(\rho_2\gamma^h\right)^{-\kappa}$$

$$= 4\rho_1 c_1 \rho_2^{-\kappa} \sum_{h=0}^{H-1} \gamma^{h(1-\kappa)} \tag{3.31}$$

$$\le \frac{4\rho_1 c_1 \rho_2^{-\kappa}}{1 - \gamma^{1-\kappa}}.$$

From (3.31), we can see that $\mathbb{E}\left[R^\pi\left(\mathcal{T}_2\right)\right]$ is upper bounded by a constant as $\mathcal{T}_2$ is a finite tree.

Step 3: Bounding the Regret on $\mathcal{T}_3$ For each node in Ω_h, its parents should be included by Φ_{h-1}. Thus, all the beams in Ω_h are $4\rho_1\gamma^{h-1}$-optimal and the cardinality of Ω_h is smaller than $2|\Phi_{h-1}|$. Besides, with the results in Lemma 3.1, we have

$$\mathbb{E}\left[N_{h,j}(t)\right] = \frac{8\sigma^2 \log t}{\left(\rho_1\gamma^h\right)^2} + c$$

for any $2\rho_1\gamma^{h-1}$-optimal nodes. Thus, the regret on $\mathcal{T}_3$ is given by

$$\mathbb{E}\left[R^\pi\left(\mathcal{T}_3\right)\right] \le \sum_{h=1}^{H} 4\rho_1\gamma^{h-1}2|\Phi_{h-1}|\mathbb{E}\left[N_{h,j}(T)\right]$$

$$\le 8\rho_1 c_1 \rho_2^{-\kappa} \sum_{h=1}^{H} \gamma^{(h-1)(1-\kappa)} \left(\frac{8\sigma^2 \log T}{\left(\rho_1\gamma^h\right)^2} + c\right). \tag{3.32}$$

Finally, substituting (3.30), (3.31), and (3.32) into (3.29), we have

$$R^{\pi}(T) \leq 4\rho_1 \gamma^H T + \frac{4\rho_1 c_1 \rho_2^{-\kappa}}{1 - \gamma^{1-\kappa}}$$

$$+ 8\rho_1 c_1 \rho_2^{-\kappa} \sum_{h=1}^{H} \gamma^{(h-1)(1-\kappa)} \left(\frac{8\sigma^2 \log T}{(\rho_1 \gamma^h)^2} + c \right) \tag{3.33}$$

$$= O\left(\gamma^H T + \log T \gamma^{-H(1+\kappa)} \right)$$

$$= O\left(T^{\frac{\kappa+1}{\kappa+2}} (\log T)^{\frac{1}{\kappa+2}} \right).$$

The last step is obtained from setting γ^H as the order of $(T/\log T)^{-1/(\kappa+2)}$ [33]. If the smoothness of the function is known, we can set $\alpha = \beta$ such that $\kappa = 0$ [33]. Hence, (3.33) can be rewritten as $O\left(\sqrt{T \log T} \right)$, and the theorem is proved. $\qquad \square$

Remark 3.4 Theorem 3.2 indicates the expected cumulative regret of HBA is sublinear in the time horizon T, i.e.,

$$\lim_{T \to \infty} \frac{R^{\pi}(T)}{T} = 0.$$

Since the regret of each slot decreases over time, the proposed algorithm is asymptotically optimal. Hence, the proposed algorithm converges to the optimal beam over time. Moreover, for finite time horizon T, the regret bound characterizes the convergence speed of the proposed algorithm.

3.6 Performance Evaluation

3.6.1 Simulation Setup

We simulate an IEEE 802.11ad system operating at 60 GHz with a bandwidth of 2.16 GHz [35, 36]. Consider an outdoor scenario, such as university campus, unless otherwise specified, the transmission distance between the transmitter and receiver is set to 20 m. The average effective isotropically radiated power (EIRP) P_e is fixed at 50 dBm,[1] which is in line with FCC regulations for 60 GHz unlicensed bands [37, 38]. Taking into account the directional antenna gain, the transmit power is $P = P_e - 10 \log_{10} N$. For instance, the transmit powers are set to around 32 dBm and 23 dBm for the 64 and 512 antenna arrays, respectively. It should be noted that the mmWave channel is sparse; therefore, we set the maximum number of channel

---

[1] For outdoor applications with the high antenna gain, the average EIRP limit is up to 82 dBm [37].

paths to 5, including one dominant LOS path and four NLOS paths. For the LOS
path, the path loss is modeled as

$$PL(\text{dB}) = 32.5 + 20\log_{10}(f) + 10\xi\log_{10}(d) + \chi \tag{3.34}$$

where f, ξ, d, and χ represent the carrier frequency, path loss exponent, transmission distance, and shadow fading, respectively. The shadow fading follows $N(0, \sigma^2)$
where σ is set to be 2 dB [39]. Note that the channel fluctuation in the simulation
is mainly caused by the shadow fading. In addition, according to practical in-field
measurements, NLOS paths suffer around 10 dB more path loss than the LOS path
[31]. We assume that the extra NLOS path loss follows a uniform distribution within
[7, 13] dB. Furthermore, for the HBA algorithm, the RSS within $[-80, -20]$ dBm
is mapped to a reward within [0, 1]. The algorithm parameters ρ_1, γ, and ζ are
set to 3, 0.5, and 0.1, respectively, based on extensive simulation trials. Important
simulation parameters are listed in Table 3.2. We evaluate the performance via
Monte-Carlo simulations. Simulation results are averaged under 50,000 samples in
different channel fading and locations. The proposed HBA algorithm is compared
to the following benchmarks:

- **Exhaustive search algorithm**: this approach scans all the combination of
  transmitter and receiver beam pair, whose search complexity is $O(N^2)$.
- **IEEE 802.11ad** [40]: In this industrial method, one side (transmitter or receiver)
  scans the beam space, while the other side keeps omni-directional. The corresponding beam measurement complexity is $O(N)$.

Table 3.2 Simulation
parameters in beam alignment

Parameter	Value
Noise spectrum density (N_o)	-174 dBm/Hz
System bandwidth (W)	2.16 GHz
Carrier frequency (f)	60 GHz
Path loss exponent (ξ)	1.74
Shadowing fading variance (σ)	2 dB
Signal range	$[-80, -20]$ dBm
SSW frame duration (T_{SSW})	15.8 us
Beacon interval duration (T_{BI})	100 ms
Number of beams (N)	{8-512}
Effective isotropically radiated power (P_e)	50 dBm
Number of paths (L)	{1-5}
Algorithm parameters (ρ_1, γ)	(3, 0.5)
Terminating condition threshold (ζ)	0.1
Time horizon (T)	1000 time slots
Extra NLOS path loss	$U(7, 13)$ dB
Transmission distance (d)	20 m

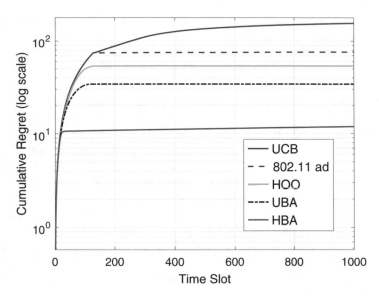

Fig. 3.6 Cumulative regret performance in the single-path channel

- **UCB** [32]: The celebrated algorithm selects the beam without exploiting both correlation structure and prior knowledge. The confidence margin is $\eta_u \sqrt{2 \log t / N_{b_i}(t)}$, where the learning rate η_u is set to be 0.2.
- **Unimodal beam alignment (UBA)** [3]: The algorithm exploits the unimodal structure among beams to perform BA. Hence, it works in a "hill-climbing" manner, which selects the best beam among the neighboring beams at each time slot.
- **HOO** [33]: The algorithm uses beam correlation to select the beam instead of the prior knowledge. The confidence margin is $\eta_h \sqrt{2 \log t / N_{h,j}(t)} + c_1 \gamma^h$, where the learning rate η_h is set to be 0.1.

3.6.2 Cumulative Regret

The proposed algorithm is compared with other benchmarks in different channels for $N = 128$. Figure 3.6 presents the cumulative regret over time in the single-path channel. It can be clearly seen that the proposed HBA algorithm significantly outperforms other benchmarks in terms of BA overhead. In addition, Fig. 3.7 shows the cumulative regret performance in two-path channels. We can get several significant observations from the simulation results. At first, HBA conspicuously outperforms other benchmarks. The observed "bounded regret" behavior is consistent with the theoretical results in Theorem 3.2. Then, HBA converges much faster than other benchmarks. Specifically, since HBA exploits both correlation structure and prior

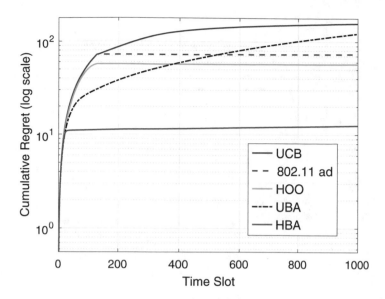

Fig. 3.7   Cumulative regret performance in the multipath channel

information to accelerate the BA process, while other benchmarks only exploit correlation structure or not, HBA only takes around 25 time slots to converge to the optimal beam. Interestingly, over time, the performance of UBA algorithm is even worse than the BA method in IEEE 802.11ad which does not utilize the correlation structure. This is because the UBA algorithm is designed based on the unimodal structure among beams, while the reward function evolves to a multimodal structure in the multipath channel. The consequence of such model mismatch is much worse than not utilizing the correlation structure at all.

We further evaluate the impact of the channel fluctuation distribution on the regret performance in Fig. 3.8. In order to evaluate the dependency of Gaussian distribution, we compared the performance of two widely adopted non-Gaussian distributions, uniform distribution and Rayleigh distribution. The performance under non-Gaussian settings is very close to that under the Gaussian distribution, which means that the proposed algorithm can be applied in various settings. Figure 3.8 also illustrates the influence of the channel fluctuation variance (σ^2). As expected, the cumulative regret increases as the variance increases, because more exploration efforts are required in highly fluctuated channels.

Finally, we evaluate the HBA algorithm with different number of paths in Fig. 3.9. Obviously, the accumulated regret only slightly grows as the number of paths increases, because more beams should be explored in the more sophisticated channel. More importantly, we prove the effectiveness and robustness of the algorithm under multipath channels.

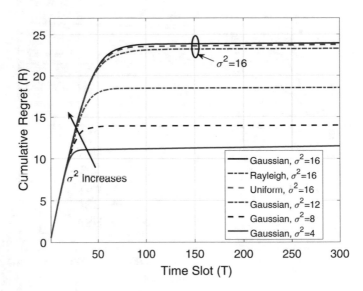

Fig. 3.8 Impact of channel distribution and variance

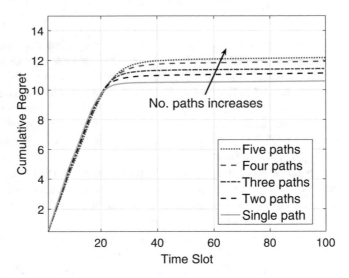

Fig. 3.9 Cumulative regret with respect to number of paths

3.6.3 Measurement Complexity and Beam Detection Accuracy

The regret performance only reflects the bounded fact of regret, not necessarily the actual performance. Next, we use the following two metrics to evaluate the performance of HBA: the number of measurements and beam detection accuracy.

We first evaluate the scalability of the proposed algorithm with the number of beams, as shown in Fig. 3.10. It is evident that the proposed algorithm significantly reduces the required number of measurements as compared to other benchmarks. For a small number ($N = 32$) of beams, the number of measurements of proposed algorithm is about 2 times less than that of the 802.11ad benchmark. In addition, the proposed algorithm achieves a higher performance gain in the case of a larger number of beams. For instance, in the case of a large number ($N = 512$) beam, the proposed algorithm only needs around 40 measurements to identify the optimal beam, while the 802.11ad benchmark requires 12 times more attempts. The reason is that the proposed algorithm only needs to explore a few beams by leveraging the correlation structure and the prior knowledge, while the BA approach in 802.11ad needs to explore all the beams. The results validate that the proposed algorithm is a scalable solution even with a large number of beams

We further study the performance in the multipath channel. Due to the inherent sparse characteristics of the mmWave channel, the number of paths ranges from 1 to 5. Firstly, the numbers of measurements in terms of the number of paths are compared in Fig. 3.11. It can be concluded that, as the number of paths increases, the number of measurements increases slightly. For example, for a 128-beam case, the number of measurements in the five-path channel increases by 15% compared to that in the single-path channel. Secondly, beam detection accuracy performance is presented in Fig. 3.12. The HBA algorithm detects the optimal beam with a high probability, even in sophisticated multipath channels. Simulation results show that even in the worst case, the beam detection accuracy can reach more than 97%. In addition, the beam detection accuracy slightly decreases as the number of paths increases. For a large number ($N = 256$) of beams, the beam detection accuracy drops from 99.6% of single-path channel to 97.4% of five-path channel due to the complex multipath channel.

Fig. 3.10 Number of beam measurements in the single-path channel. Error bars show the 90 percentile performance

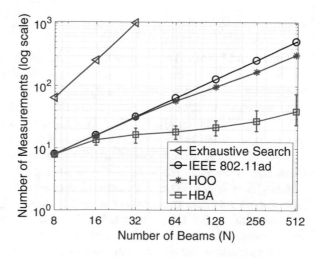

Fig. 3.11 Number of beam
measurements in the
multipath channel

Fig. 3.11 Number of beam
measurements in the
multipath channel

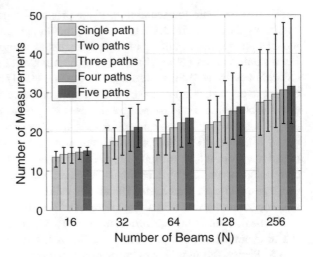

Fig. 3.12 Beam detection
accuracy in the multipath
channel

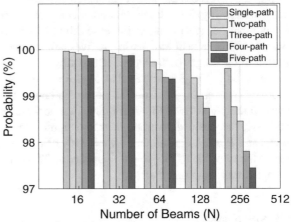

Figures 3.13 and 3.14 illustrate the impact of the transmission distance on the performance. We first observe that the number of measurements increases in terms of the transmission distance, as shown in Fig. 3.13. Specifically, the number of measurements increases by 32% as distance increases from 5 m to 50 m for $N = 128$. This is because the RSS is weaker for a longer distance such that limited information can be extracted from nearby beams. Hence, the proposed algorithm needs to explore more beams to identify the optimal beam for remote users. Even for remote users, the proposed BA algorithm performs better than the 802.11ad benchmark. When the distance increases to 50 m, our algorithm requires approximately 44 measurements for $N = 256$, which still reduces the number of measurements by 5.8 times compared with the 802.11ad benchmark. Finally, the beam detection accuracy is presented in Fig. 3.14. Even in the low SNR case, the proposed algorithm can detect the optimal beam with a high probability.

Fig. 3.13 Number of measurements performance comparison with respect to transmission distance in two-path channels

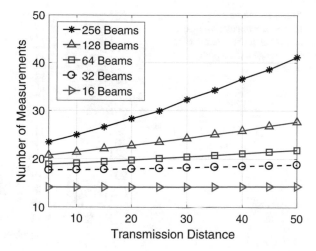

Fig. 3.14 Beam detection accuracy performance comparison with respect to transmission distance in two-path channels

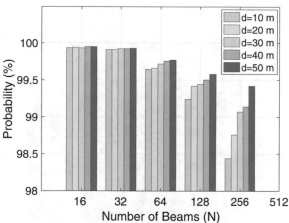

For implementation consideration, Figs. 3.15 and 3.16 present the performance of HBA under coarse prior knowledge conditions. Define the ratio of estimated variance (σ_e^2) to accurate variance as a measure of the coarse prior knowledge, i.e.,

$$\eta = \frac{\sigma_e^2}{\sigma^2}.$$

Therefore, the coarse prior knowledge can be divided into two categories: the underestimated prior knowledge when $\eta < 1$ and the overestimated prior knowledge when $\eta > 1$. We can observe from Fig. 3.15 that the number of measurements increases as η increases from 0.25 to 4. Specifically, for a 256-beam case, the HBA algorithm with the overestimated prior knowledge for $\eta = 4$ requires more beam measurements than the HBA algorithm with accurate prior knowledge. In order to adapt to the uncertainty of the reward, overestimating the prior knowledge

Fig. 3.15 Number of measurement performance comparison with coarse prior knowledge in two-path channels

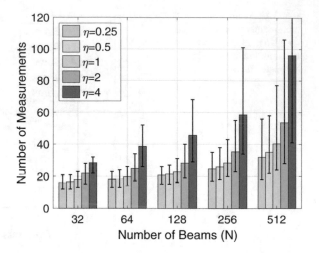

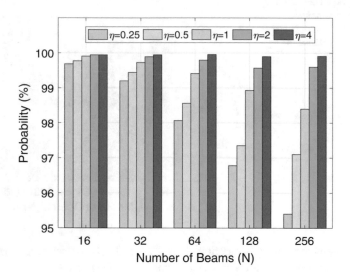

Fig. 3.16 Beam detection accuracy performance comparison with coarse prior knowledge in two-path channels

leads to a larger confidence margin, which requires more exploration efforts to obtain better beam detection accuracy, as shown in Fig. 3.16. On the contrary, when prior knowledge is underestimated, the number of measurements is slightly smaller than that with accurate prior knowledge, while the beam detection accuracy reduces due to insufficient exploration efforts. More importantly, even with the coarse prior knowledge, the proposed algorithm can substantially reduce the number of measurements compared with benchmarks and achieve high beam detection accuracy. For a 256-beam case, even in the worst case, the proposed algorithm reduces the number of measurements by 6 times in comparison with the BA method in 802.11ad.

3.6.4 Beam Alignment Latency

Practical BA latency needs to take the 802.11ad protocol into consideration, which is different from a simple product of the number of measurements and the duration of each measurement. In the protocol, BA must be carried out in the associated beamforming training (A-BFT) stage, which contains eight A-BFT slots, and each A-BFT slot contains 16 sector sweep (SSW) frames. Each SSW frame can only provide one measurement for one beam, and the duration is about 15.8 us [40]. If the BA process cannot be completed in the A-BFT stage of the current beacon interval (BI), this BA process has to wait for the A-BFT stage in the next BI, which increases the BA delay of the entire BI duration. In the simulation, the duration of BI is set to 100 ms [40]. With the above protocol, BA latency is calculated based on the average number of measurements. Table 3.3 presents the BA latency with different numbers of beams in the two-path channel. As expected, the BA latency increases as the number of beams increases. For the case of only one user, compared to the BA method in 802.11ad, the proposed algorithm reduces the BA latency significantly. In particular, for a large number ($N = 256$) of beams, the BA latency drops from 106.07 ms to only 0.88 ms. This is because the BA process of the proposed algorithm requires only a small amount of measurements to identify the optimal beam and thus can be completed in one BI. Furthermore, we can observe a larger performance gain in the four-user case. In contrast to the BA method in 802.11ad which incurs more than 700 ms latency for a 256-beam-phase array, the proposed algorithm takes about 2.21 ms, which corresponds to two orders of magnitude gain.

Table 3.3 BA latency comparison in the multipath channel

N	One user		Four-user	
	802.11ad	HBA	802.11ad	HBA
8	0.25 ms	0.25 ms	0.63 ms	0.63 ms
16	0.51 ms	0.45 ms	1.26 ms	1.12 ms
32	1.01 ms	0.56 ms	2.53 ms	1.39 ms
64	2.02 ms	0.61 ms	103.03 ms	1.53 ms
128	4.04 ms	0.71 ms	304.04 ms	1.78 ms
256	106.07 ms	0.88 ms	706.07 ms	2.21 ms

3.7 Summary

In this chapter, we have investigated the BA problem in mmWave systems to find the optimal beam pair. We developed HBA, a learning algorithm which uses the inherent correlation structure among beams and the prior knowledge on the channel fluctuation to speed up the BA process. The proposed HBA algorithm can utilize a small number of beam measurements to identify the optimal beam with a high probability, even in the case of a large number of beams.

References

1. W. Wu, N. Cheng, N. Zhang, P. Yang, W. Zhuang, X. Shen, Fast mmwave beam alignment via correlated bandit learning. IEEE Trans. Wireless Commun. **18**(12), 5894–5908 (2019)
2. J. Qiao, Y. He, X. Shen, Proactive caching for mobile video streaming in millimeter wave 5G networks. IEEE Trans. Wireless Commun. **15**(10), 7187–7198 (2016)
3. M. Hashemi, A. Sabharwal, C.E. Koksal, N.B. Shroff, Efficient beam alignment in millimeter wave systems using contextual bandits, in *Proc. IEEE INFOCOM* (2018), pp. 2393–2401
4. H. Hassanieh, O. Abari, M. Rodriguez, M. Abdelghany, D. Katabi, P. Indyk, Fast millimeter wave beam alignment, in *Proc. ACM SIGCOMM* (2018), pp. 432–445
5. Z. Marzi, D. Ramasamy, U. Madhow, Compressive channel estimation and tracking for large arrays in mm-Wave picocells. IEEE J. Sel. Topics Signal Process. **10**(3), 514–527 (2016)
6. S. Sur, I. Pefkianakis, X. Zhang, K.H. Kim, WiFi-assisted 60 GHz wireless networks, in *Proc. ACM MOBICOM* (2017), pp. 28–41
7. P. Zhou, X. Fang, Y. Fang, Y. Long, R. He, X. Han, Enhanced random access and beam training for millimeter wave wireless local networks with high user density. IEEE Trans. Wireless Commun. **16**(12), 7760–7773 (2017)
8. J. Wang, Z. Lan, C. Pyo, T. Baykas, C. Sum, M.A. Rahman, J. Gao, R. Funada, F. Kojima, H. Harada, S. Kato, Beam codebook based beamforming protocol for multi-Gbps millimeter-wave WPAN systems. IEEE J. Sel. Areas Commun. **27**(8), 1390–1399 (2009)
9. Z. Xiao, T. He, P. Xia, X.-G. Xia, Hierarchical codebook design for beamforming training in millimeter-wave communication. IEEE Trans. Wireless Commun. **15**(5), 3380–3392 (2016)
10. X. Sun, C. Qi, G.Y. Li, Beam training and allocation for multiuser millimeter wave massive MIMO systems. IEEE Trans. Wireless Commun. **18**(2), 1041–1053 (2019)
11. A. Ali, N. González-Prelcic, R.W. Heath, Millimeter wave beam-selection using out-of-band spatial information. IEEE Trans. Wireless Commun. **17**(2), 1038–1052 (2018)
12. M. Hashemi, C.E. Koksal, N.B. Shroff, Out-of-band millimeter wave beamforming and communications to achieve low latency and high energy efficiency in 5G systems. IEEE Trans. Commun. **66**(2), 875–888 (2018)
13. Y. Shabara, C.E. Koksal, E. Ekici, Linear block coding for efficient beam discovery in millimeter wave communication networks, in *Proc. IEEE INFOCOM* (2018), pp. 2285–2293
14. X. Shen, J. Gao, W. Wu, K. Lyu, M. Li, W. Zhuang, X. Li, J. Rao, AI-assisted network-slicing based next-generation wireless networks. IEEE Open J. Veh. Technol. **1**(1), 45–66 (2020)
15. X. You et al., Towards 6G wireless communication networks: Vision, enabling technologies, and new paradigm shifts. Sci. China Inf. Sci. **64**(1), 1–74 (2021)
16. W. Zhuang, Q. Ye, F. Lyu, N. Cheng, J. Ren, SDN/NFV-empowered future IoV with enhanced communication, computing, and caching. Proc. IEEE **108**(2), 274–291 (2019)
17. W. Wu, N. Chen, C. Zhou, M. Li, X. Shen, W. Zhuang, X. Li, Dynamic RAN slicing for service-oriented vehicular networks via constrained learning. IEEE J. Sel. Areas Commun. **39**(7), 2076–2089 (2021)

18. K. Qu, W. Zhuang, Q. Ye, X. Shen, X. Li, J. Rao, Dynamic flow migration for embedded services in SDN/NFV-enabled 5G core networks. IEEE Trans. Commun. **68**(4), 2394–2408 (2020)
19. W. Wu, P. Yang, W. Zhang, C. Zhou, X. Shen, Accuracy-guaranteed collaborative DNN inference in industrial IoT via deep reinforcement learning. IEEE Trans. Ind. Informat. **17**(7), 4988–4998 (2021)
20. Z. Wang, C. Shen, Small cell transmit power assignment based on correlated bandit learning. IEEE J. Sel. Areas Commun. **35**(5), 1030–1045 (2017)
21. C. Shen, R. Zhou, C. Tekin, M. van der Schaar, Generalized global bandit and its application in cellular coverage optimization. IEEE J. Sel. Topics Signal Process. **12**(1), 218–232 (2018)
22. P. Yang, N. Zhang, S. Zhang, L. Yu, J. Zhang, X. Shen, Content popularity prediction towards location-aware mobile edge caching. IEEE Trans. Multimedia **21**(4), 915–929 (2019)
23. S. Müller, O. Atan, M. van der Schaar, A. Klein, Context-aware proactive content caching with service differentiation in wireless networks. IEEE Trans. Wireless Commun. **16**(2), 1024–1036 (2017)
24. P. Yang, N. Zhang, S. Zhang, K. Yang, L. Yu, X. Shen, Identifying the most valuable workers in fog-assisted spatial crowdsourcing. IEEE Internet Things J. **4**(5), 1193–1203 (2017)
25. Y. Sun, S. Zhou, J. Xu, EMM: Energy-aware mobility management for mobile edge computing in ultra dense networks. IEEE J. Sel. Areas Commun. **35**(11), 2637–2646 (2017)
26. N. Gulati, K.R. Dandekar, Learning state selection for reconfigurable antennas: A multi-armed bandit approach. IEEE Trans. Antennas Propag. **62**(3), 1027–1038 (2014)
27. G.H. Sim, S. Klos, A. Asadi, A. Klein, M. Hollick, An online context-aware machine learning algorithm for 5G mmWave vehicular communications. IEEE/ACM Trans. Netw. **26**(6), 2487–2500 (2018)
28. I. Chafaa, E.V. Belmega, M. Debbah, Adversarial multi-armed bandit for mmwave beam alignment with one-bit feedback, in *Proc. ACM ValueTools* (2019)
29. W. Wu, Q. Shen, M. Wang, X. Shen, Performance analysis of IEEE 802.11.ad downlink hybrid beamforming, in *Proc. IEEE ICC* (2017)
30. M.R. Akdeniz, Y. Liu, S. Sun, S. Rangan, T.S. Rappaport, E. Erkip, Millimeter wave channel modeling and cellular capacity evaluation. IEEE J. Sel. Areas Commun. **32**(6), 1164–1179 (2013)
31. A. Maltsev, R. Maslennikov, A. Sevastyanov, A. Khoryaev, A. Lomayev, Experimental investigations of 60 GHz WLAN systems in office environment. IEEE J. Sel. Areas Commun. **27**(8), 1488–1499 (2009)
32. P. Auer, N. Cesa-Bianchi, P. Fischer, Finite-time analysis of the multiarmed bandit problem. Mach. Learn. **47**(2), 235–256 (2002)
33. S. Bubeck, G. Stoltz, C. Szepesvári, R. Munos, Online optimization in X-armed bandits, in *Proc. NIPS* (2009)
34. P.B. Reverdy, V. Srivastava, N.E. Leonard, Modeling human decision making in generalized Gaussian multiarmed bandits. Proc. IEEE **102**(4), 544–571 (2014)
35. W. Wu, N. Zhang, N. Cheng, Y. Tang, K. Aldubaikhy, X. Shen, Beef up mmwave dense cellular networks with D2D-assisted cooperative edge caching. IEEE Trans. Veh. Technol. **68**(4), 3890–3904 (2019)
36. W. Wu, N. Cheng, N. Zhang, P. Yang, K. Aldubaikhy, X. Shen, Performance analysis and enhancement of beamforming training in 802.11ad. IEEE Trans. Veh. Technol. **69**(5), 5293–5306 (2020)
37. FCC, Report and order and further notice of proposed rulemaking, federal communications commission (2016)
38. J. Du, R.A. Valenzuela, How much spectrum is too much in millimeter wave wireless access. IEEE J. Sel. Areas Commun. **35**(7), 1444–1458 (2017)
39. 3GPP, Technical specification group radio access network: Study on channel model for frequencies from 0.5 to 100 GHz (2017)
40. IEEE Standards, IEEE standards 802.11 ad-2012: Enhancement for very high throughput in the 60 GHz band (2012)

Chapter 4
Beamforming Training Protocol Design and Analysis

4.1 Introduction

Millimeter-wave (mmWave) band communication, particularly at the unlicensed 60 GHz frequency band, has received considerable attention due to its application in short-range indoor scenarios, such as wireless personal area networks (WPANs) and wireless local area networks (WLANs). For example, the IEEE 802.15.3c standard is ratified for mmWave communication in WPANs, and the IEEE 802.11ad standard is ratified for mmWave communication in WLANs. Both of them operate at the unlicensed 60 GHz frequency band. Although mmWave communication can offer high data rate transmission, it suffers from severe free-space path loss due to operating at a high frequency band [1–3]. *Beamforming* technology which focuses the radio-frequency power in a narrow direction is adopted at both the transmitter and receiver to compensate for the huge path loss in mmWave communication. As reliable communication is only possible when the beamforming of both the transmitter and receiver is properly aligned, a *beamforming training* process between the transmitter and receiver is needed. Without the beamforming training process, the data rate of mmWave communications would drop from several Gbps to only a few hundred Mbps [4, 5]. Therefore, designing an efficient beamforming training scheme is significant for mmWave communication.

In the literature, a few recent works have investigated beamforming training schemes, e.g., codebook-based beam search [6], compressed sensing schemes [7], and out-of-band solutions [8]. With recent trend of leveraging advanced machine learning methods to address wireless networking problems [9, 10], some machine learning-based solutions are also developed for low-complexity beamforming training or beam alignment [11, 12]. Even though the existing works can greatly enhance beamforming training performance, they focus on investigating the performance from a perspective of the physical layer. However, the medium access control (MAC) layer is also very important for beamforming training performance. The contention feature of the MAC layer is seldom considered in previous works, i.e., the multiple stations (STAs) compete for the same beamforming training time.[1] Even with efficient beamforming training schemes, a coarse MAC protocol would result in severe collisions for beamforming training, which wastes the cherished beamforming training time and incurs extra beamforming training latency. Hence, the elaborate analysis and tailored enhancement of the MAC protocol for beamforming training are of paramount importance.

IEEE 802.11ad specifies a new distributed beamforming training MAC protocol, namely, BFT-MAC protocol, which can coordinate beamforming training among multiple STAs. Specifically, the duration of beamforming training is divided into multiple associated beamforming training (A-BFT) slots. All active STAs in the coverage contend for these A-BFT slots in a contention and backoff manner in order to obtain a beamforming training opportunity. However, the performance of BFT-MAC in dense user scenarios is still unclear. This is due to the following two reasons: (1) due to the "deafness" problem caused by beamforming (i.e., directional antennas), i.e., an STA may not sense the transmission of other STAs. As such, the BFT-MAC protocol is different from the traditional carrier sensing-based MAC protocols in microwave WLANs. Hence, existing analytical models for traditional microwave WLANs cannot applied for the BFT-MAC protocol. (2) Previous works in [13] investigate and simulate the MAC performance with a finite number of STAs, which can hardly provide theoretical insights for dense user scenarios. Therefore, we argue that a thorough new analytical model for the BFT-MAC protocol is necessary and significant. Furthermore, only at most eight A-BFT slots are provided in 802.11ad standard, and the collision probability is extremely high in dense user scenarios due to the lack of A-BFT slots. The severe collision results in low throughput and high beamforming training latency. Thus, it is required to design an enhancement scheme on 802.11ad to improve the performance in dense user scenarios. The aforementioned MAC performance analysis and enhancement are designed for the existing 802.11ad standard. Hence, the *first issue* focuses on improving the beamforming training efficiency for *single-user transmission*, i.e., only one user transmits data at one time.

In addition to single-user transmission, to further enhance date rates, multiuser transmission schemes that enable the concurrent transmission of multiple users are desired. Multiuser transmission is expected to be adopted in next-generation mmWave WLAN, i.e., 802.11ay. To achieve multiuser beamforming in mmWave networks, hybrid beamforming is considered as a low-complexity strategy [14]. Specifically, hybrid beamforming consists of an analog beamforming part and a digital beamforming part. The analog beamforming part aims to provide directional antenna gain by controlling the transmitted signal phase at each antenna, while the digital beamforming part aims to mitigate multiuser inference by judiciously designing a proper baseband beamforming matrix. Theoretical results have proved that hybrid beamforming can achieve close-to-optimal performance as compared with fully digital beamforming while significantly reducing implementation complex-

---

[1] In this chapter, we use the word "STA" and "user" interchangeably.

ity [15]. Thus, designing an 802.11ad-compliant *multiuser beamforming protocol* based on hybrid beamforming methods and analyzing the overhead of the designed protocol are very important for further improving data rate for future mmWave networks. The *second issue* focuses on designing and analyzing the beamforming training protocol for enabling *multiuser transmission*.

In this chapter, we focus on addressing the above two issues. Specifically, this chapter can be divided into two parts. The first part is for the *single-user transmission* scenario, which is to analyze and enhance beamforming training performance from the perspective of the MAC protocol. The second part is for the *multiuser transmission* scenario, which is to design a novel multiuser beamforming training protocol and analyze its performance from the perspective of supporting multiuser transmission.

In the *first part*, we focus on the MAC performance analysis and enchantment for the current 802.11ad BFT-MAC. We try to answer the following two questions: *(1) How good is the performance of BFT-MAC? (2) How can one further enhance the performance of BFT-MAC in dense user scenarios?* Firstly, a two-dimensional Markov chain-based analytical model, which models the number of consecutive collisions and the backoff time as a state, is presented to evaluate the BFT-MAC performance. The presented analytical model can unveil the relationship among the number of A-BFT slots, the number of STAs, and MAC parameters on the BFT-MAC performance. Secondly, given the analytical model, the closed-form expressions of the normalized throughput and beamforming training latency are derived theoretically, respectively. Moreover, asymptotic analysis in dense user scenarios indicates that the normalized throughput depends on the ratio between the number of STAs and the number of A-BFT slots. Particularly, theoretical analysis indicates that maximum normalized throughput is barely $1/e$, which is the same as the slotted ALOHA protocol. Thirdly, since the performance substantially degrades in dense user scenarios due to the limitation of A-BFT slots in practical mmWave networks, we discuss an *enhancement scheme* which adaptively adjusts MAC parameters according to the user density to improve MAC performance. Extensive simulation results demonstrate that the enhancement scheme can significantly improve the normalized throughput and reduce the beamforming training latency, as compared to 802.11ad with the default protocol parameter configuration.

In the *second part*, we focus on the multiuser beamforming protocol design and analysis. We target at answering the following question: *How to design a multiuser beamforming training protocol and analyze its performance?* Specifically, a multiuser beamforming training protocol is first introduced to support the hybrid beamforming algorithm based on 802.11ad. Second, we analyze the overhead of the introduced multiuser beamforming training protocol as well as the throughput gain. Particularly, theoretical analysis demonstrates that beamforming training overhead increases linearly with the number of users, which can diminish throughput gain when the number of users is large. Extensive simulations are conducted to validate the effectiveness of the introduced multiuser beamforming training protocol. Simulation results show that the throughput gain increases and then decreases with the increase of the number of users. Thus, our observation suggests that there exists an optimal number of users that hybrid beamforming-enabled mmWave networks should support.

The remainder of this chapter is organized as follows. For the *first part*, an overview of beamforming training schemes is presented in Sect. 4.2. Then, we show beamforming training protocol in 802.11ad from a perspective of the MAC layer in Sect. 4.3. The presented analytical model, the corresponding performance analysis, and the introduced enhancement scheme are given in Sect. 4.4. In Sect. 4.5, extensive simulations are conducted to validate the presented analytical model and the enhancement scheme. For the *second part*, we design a multiuser beamforming training protocol based on hybrid beamforming algorithms and analyze its beamforming training overhead in Sect. 4.6. Simulations results are provided to demonstrate the presented protocol's performance in Sect. 4.7. Finally, Sect. 4.8 concludes this chapter.

4.2 Existing Works on Beamforming Training

In the following, we present the existing works on beamforming training schemes and MAC performance analysis of beamforming training protocol in Sects. 4.2.1 and 4.2.2, respectively.

4.2.1 Beamforming Training Schemes

In the literature, there are a large amount of efforts on developing efficient beamforming training schemes in mmWave networks. The authors proposed a codebook-based search scheme in [6]. In the proposed scheme, the beamwidth of beamforming is adjusted in each step until the optimal beam is identified. To support multiuser transmission, a low-complexity hybrid beamforming algorithm is developed in [16]. The hybrid beamforming algorithm is a combination of analog beamforming and digital beamforming, which can reduce implementation complexity. By utilizing the sparse characteristic of mmWave channels, the authors developed a compressed sensing-based beamforming training method with a low complexity [7]. Some out-of-band schemes were developed in [8], which can exploit traditional Wi-Fi signals to reduce beamforming training overhead. In another direction, by utilizing the multi-armed beam feature of directional antennas, the authors proposed a fast beamforming training scheme [17]. While the above works can enhance the beamforming training efficiency, these works do not consider the contentions of multiple STAs in the beamforming training protocol. Different from these works, our work in this chapter concentrates on studying beamforming training performance from a perspective of the MAC layer.

4.2.2   MAC Performance Analysis

In the literature, the *MAC performance* of different protocols has been widely investigated. The MAC protocols in traditional microwave networks have been widely analyzed in various scenarios based on the celebrated Bianchi's model [18], such as highly mobility vehicular networks [19, 20], mobile ad hoc networks [21], Internet-of-things networks [22], and wireless body area networks (WBANs) [23]. A pioneering three-dimensional Markov chain analytical model is proposed to investigate the 802.11 distributed coordinate function (DCF) performance in the drive-through vehicular networks [19]. On this basis, taking practical access procedures into consideration, an extended work investigated the throughput performance of the drive-through vehicular networks [20]. In addition, taking the impact of interference into consideration, the authors developed an analytical model for analyzing MAC performance in WBANs [23]. Although these works in [18–23] show insightful lights on the MAC performance analysis in microwave systems, they focus on the omni-directional system. Due to the adoption of beamforming technology, 802.11ad WLANs are directional communication systems. Hence, the legacy of MAC analytical models in omni-directional systems cannot be applied.

Several recent works devoted to analyzing MAC performance in *directional 802.11ad WLANs*. The authors analyzed the impact of the number of sectors in the directional transmission in [24]. Another work proposed a directional cooperative MAC protocol and analyzed its performance in [25]. However, the above works in [24, 25] focus on the MAC performance in the data transmission stage. In contrast, the MAC performance in the beamforming training stage is seldom considered, which is the bottleneck of the whole system. The following work in the first part focuses on the MAC performance in the beamforming training stage.

Some works in [13, 26–29] focus on improving the MAC performance in the *beamforming training stage*. A pioneering work [13] proposed a secondary backoff scheme in the A-BFT stage to alleviate beamforming training collisions in dense user scenarios. In this work, each STA selects not only a backoff A-BFT slot but also a secondary backoff time within the A-BFT slot. With this method, transmission collisions can be reduced. The authors leveraged the channel sparsity in the mmWave channel to develop a compressed sensing method to perform beamforming training simultaneously for a group of STAs in [26]. In addition to these works, multiple standardization efforts have been devoted to enhancing the MAC performance in the beamforming training stage. The authors in [27] spread out the access attempt over time. As such, the high collision issue can be addressed in dense user networks. Another standardization draft in [28] allowed beamforming training simultaneously for different STAs over multiple channels, thereby enhancing beamforming training efficiency. But this scheme may increase the protocol signaling overhead. In another draft, a short sector sweep (SSW) frame structure is proposed in [29]. The short SSW has a shorter packet length, as compared to a traditional SSW frame, which can increase the beamforming training capability in the 802.11ad standard since more short SSW frames can be included in an A-BFT slot. While these works can provide efficient solutions on enhancing MAC performance in the beamforming training stage, they more or less need to modify the MAC protocol. Hence, these works may be incompatible with the current 802.11ad standard.

In contrast, the following work in Sects. 4.3 and 4.4 in the first part focuses on an in-depth understanding of the 802.11ad MAC protocol for beamforming training instead of proposing new MAC protocols with distinguished features. The underlying reason is twofold. Firstly, as the most practical and adopted standard in mmWave WLANs, the 802.11ad standard is widely used in many commercial off-the-shelf (COTS) devices. Secondly, the future 802.11ay standard is highly envisioned to adopt a similar MAC protocol for beamforming training as 802.11ad. Specifically, 802.11ay would have an increased number of A-BFT slots [30, 31]. Hence, the performance of 802.11ay can be analyzed based on the proposed analytical model with some customized modification.

4.3   Beamforming Training Protocol in 802.11ad

In the following, we first present the beamforming training procedure in the 802.11ad standard in Sect. 4.3.1, and then we introduce the BFT-MAC protocol in detail in Sect. 4.3.2.

4.3.1   Beamforming Training Procedure

A WLAN compliant with the 802.11ad standard is considered in this work, which consists of an access point (AP) and multiple STAs. The AP is in charge of coordinating beamforming training, link scheduling, and network synchronization. Both the AP and STAs adopt the directional multi-gigabit mode, which means that each node is equipped with an electrically steerable directional antenna to support the directional transmission.

The *beamforming training procedure* is to establish reliable mmWave communication links between the AP and STAs. To this end, at the beginning of each beacon interval (BI), an STA needs to perform beamforming training with the AP. As illustrated in Fig. 4.1, the transmission time is divided into multiple BIs. Each BI is further segregated into:

- The beacon transmission interval (BTI) stage, during which AP performs the beamforming training with STA.
- The A-BFT stage, during which all STAs contend for beamforming training with AP.
- The announcement transmission interval (ATI) stage, which coordinates the transmission scheduling in data transmission interval (DTI).

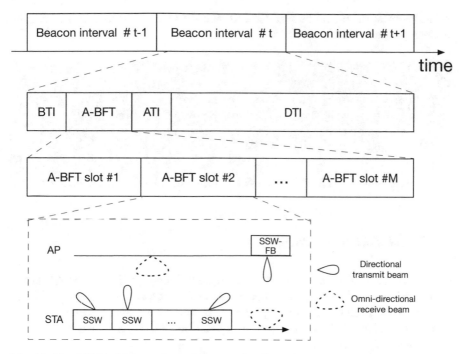

Fig. 4.1 The 802.11ad beacon interval format. An illustrative example of the beamforming training procedure is provided in the dashed square area

- The DTI stage, which facilitates directional data transmission [32]. Specifically, the DTI consists of multiple service periods (SPs) and contention-based access periods (CBAPs), where SPs are scheduled access periods and CBAPs are enhanced distributed channel access periods.

For more details, one can refer to our detailed introduction of the 802.11ad beamforming training protocol in Sect. 2.3 in Chap. 2.

This work focuses on the beamforming training in the A-BFT stage. Next, we briefly introduce the *A-BFT stage*. The A-BFT stage is further divided into multiple A-BFT slots. The A-BFT slots can provide separated beamforming training for different STAs. The detailed beamforming training procedure between the AP and an STA in an A-BFT slot is illustrated in Fig. 4.1. Specifically, an STA transmits multiple SSW frames via different directional beams, and then the AP receives these SSW frames via its omni-directional beam. The AP can identify the best transmit beam of the STA based on the received signal strength of these directional beams. In the following, an SSW-feedback (SSW-FB) frame is sent from the AP to the STA for the acknowledgment of a *successful beamforming training*. It is worth noting that an A-BFT slot can only provide one beamforming training opportunity for an STA. When two STAs compete for the same A-BFT slot, a *collision* occurs. In this case, no SSW-FB frame would be sent to STAs, and then the occurrence of the collision is aware by the transmitting STAs.

> **Remark 4.1** It is worth noting that the abovementioned beamforming training in A-BFT is a part of the entire beamforming training procedure in 802.11ad. In this chapter, we focus on the performance of beamforming training from a perspective of the MAC layer. The detailed beamforming training procedure is beyond the scope of our study. For more details, one can refer to a detailed survey in [33].

4.3.2 BFT-MAC Protocol

The 802.11ad specifies a distributed beamforming training MAC protocol, namely, *BFT-MAC protocol*, to coordinate beamforming training among multiple STAs. In the BFT-MAC protocol, each STA performs beamforming training with the AP in a contention and backoff manner. Specifically, the BFT-MAC protocol consists of the following two parts:

- *Before beamforming training*, a random A-BFT slot is selected from the range $[1, M]$ by each *active STA* for the transmission of beamforming training packets (referred to as a transmission for short hereinafter). Here, the number of A-BFT slots is denoted by M. An event of successful beamforming training depends on whether an A-BFT slot is selected by multiple STAs. If an A-BFT slot is only selected by one STA, the transmission would succeed, and the AP would send an SSW-FB frame to the STA. As the example shown in Fig. 4.2, the beamforming training attempt of *STA A* in the A-BFT slot #1 is successful. This is because this A-BFT slot is not selected by the other STAs. Otherwise, when multiple STAs select the same A-BFT slot, a collision would occur, and the AP would not send the SSW-FB frames to STAs. For example, *STA B* and *STA C* encounter a collision in beamforming training, as shown in Fig. 4.2.
- *After beamforming training*, when the number of consecutive collisions that an STA has experienced exceeds retry limit R (referred to as *dot11RSSRetryLimit* in 802.11ad), a discrete backoff time w from the range $[0, W)$ would be selected by the STA in a uniform manner. Here, the contention window size (referred to as *dot11RSSBackoff* in 802.11ad) is denoted by W. Specifically, a consecutive collision counter (referred to as *FailedRSSAttempts* in 802.11ad) is maintained by each STA. The counter indicates the number of consecutive collisions that the STA has experienced in the A-BFT stage. The consecutive collision counter is incremented by one once a collision occurs. Otherwise, the consecutive collision counter is cleared to zero upon a successful transmission.

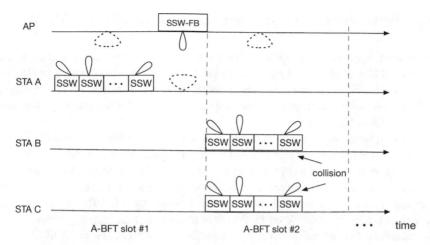

Fig. 4.2 An illustrative example of the BFT-MAC protocol. In the figure, the beamforming training in the A-BFT slots #1 is successful, while that in A-BFT slot #2 is unsuccessful since this A-BFT slot is selected by two STAs simultaneously

It is worth noting that STAs can transmit only when the backoff time is zero. The backoff time would be decremented by one at the end of one BI. Let w represent the backoff time of an STA. It means that the STA has to be frozen from transmission in the subsequent w BIs. Due to the backoff mechanism, we know that not all STAs are contending for A-BFT slots. Here, STAs whose backoff times are zero are referred to as *active STAs*, while other STAs whose backoff times are nonzero are referred to as *inactive STAs*. In this chapter, we assume that transmission is always successful unless a collision occurs for simplicity.

> *Remark 4.2* The advantages of the BFT-MAC protocol are salient, which consist of the following two:
>
> - BFT-MAC is fully distributed which is scalable with the network size.
> - The MAC protocol is simple which can be easily implemented under different scenarios.
>
> However, compared with the celebrated DCF protocol in traditional omnidirectional WLANs, *the absence of carrier sensing mechanism makes BFT-MAC susceptible to collide*, especially in dense user scenarios.

In what follows, the performance of BFT-MAC is analyzed.

4.4 Performance Analysis and Enhancement for BFT-MAC

In the following, we first propose a two-dimensional Markov chain-based analytical model for 802.11ad BFT-MAC in Sect. 4.4.1. Based on the analytical model, we analyze the BFT-MAC performance in terms of average successful beamforming training probability, average beamforming training latency, and network throughput, respectively, in Sect. 4.4.2.

The following *key assumptions* are adopted in the following analysis. Firstly, we assume a fixed number of STAs and perfect physical channel conditions (i.e., no transmission errors) in the considered network. Secondly, we consider the *saturation condition*. In saturation condition, each STA always needs to perform beamforming training at each BI for continuous data transmission [13, 18, 19]. This assumption is reasonable in the considered mmWave networks. The underlying reason is that the established communication connections are intermittent and short-lived due to the user mobility and potential blockage events. As such, the beamforming training procedure would be invoked persistently. For better illustration, a summary of important notations is listed in Table 4.1.

4.4.1 Analytical Model for BFT-MAC

The BFT-MAC protocol operates in a discrete time slotted manner. Let t denote the index of BIs. We examine a *tagged STA* to evaluate the performance of BFT-

Table 4.1 Summary of notations in the BFT-MAC protocol

Notation	Description
p	Conditional collision probability
p_s	Conditional successful transmission probability
τ	The probability that an STA is active
$C(t)$	Consecutive collision counter at time t
$B(t)$	Backoff time at time t
(r, w)	State with r consecutive collisions and w backoff time
M	Number of A-BFT slots
R	Value of retry limit
D	Average beamforming training latency
W	Contention window size
S	Normalized throughput
π	Steady-state probability vector
T_{BI}	Duration of a beacon interval
T_{SSW}	Duration of a sector sweep frame
$\mathbb{Z}^+$	Positive integer set
F	Number of SSW frames in an A-BFT slot

MAC. The status of the tagged STA is represented by a two-dimensional Markov chain $\{C(t), B(t)\}$. Here, $C(t) \in [0, R]$ represents the number of consecutive collisions that the tagged STA has experienced, and $B(t) \in [0, W - 1]$ represents the current backoff time of the tagged STA. For instance, state (r, w) means that the tagged STA has experienced r consecutive collisions and its current backoff time is w. Let p denote the average collision probability of the tagged STA. It is worth noting that p is the *conditional collision probability*. This is because the collision occurs only when the tagged STA is active. Hence, $1 - p$ denotes the average conditional successful transmission probability. We depict the state transition diagram in Fig. 4.3.

The state transition is governed by the following *events*, and the corresponding *one-step transition probabilities* are given as follows.

- *Transmission collision:* The consecutive collision counter is incremented by one upon a transmission collision, when the counter does not exceed the retry limit. In this case, the STA transits from state $(r, 0)$ to state $(r + 1, 0)$, and the corresponding transition probability is given by

$$\mathbb{P}(r + 1, 0 | r, 0) = p, \forall r \in [0, R - 2]. \tag{4.1}$$

- *Successful transmission:* The consecutive collision counter is cleared to zero upon a successful transmission. In this case, the STA transits from state $(r, 0)$ to state $(0, 0)$ according to the following transition probability

$$\mathbb{P}(0, 0 | r, 0) = 1 - p, \forall r \in [0, R]. \tag{4.2}$$

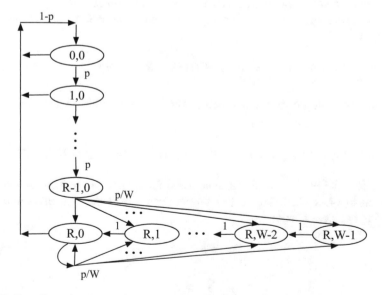

Fig. 4.3 Two-dimensional Markov chain-based analytical model for BFT-MAC

- *Backoff time selection:* The STA selects a random backoff time from interval $w \in [0, W-1]$ when the consecutive collision counter reaches retry limit R. In this case, the STA transits from state $(R-1, 0)$ to backoff state (R, w), and the corresponding transition probability is given by

$$\mathbb{P}\left(R, w | R-1, 0\right) = \frac{p}{W}, \forall w \in [0, W-1]. \tag{4.3}$$

In the subsequent w BIs, the STA would be frozen from transmission in the A-BFT stage.

- *Backoff in the frozen states:* In the frozen states, the backoff time is decremented by one after every BI. In this case, an STA transits from state (R, w) to state $(R, w-1)$. The one step transition probability is given by

$$\mathbb{P}\left(R, w-1 | R, w\right) = 1, \forall w \in [1, W-1]. \tag{4.4}$$

- *Transmission collision when the retry limit is reached:* When the consecutive collision counter reaches the retry limit, it would not be incremented. In this case, once a transmission collision occurs, an STA transits from state $(R, 0)$ to state (R, w). The corresponding transition probability is given by

$$\mathbb{P}\left(R, w | R, 0\right) = \frac{p}{W}, \forall w \in [0, W-1]. \tag{4.5}$$

Then, the tagged STA has to stay in these backoff states in the subsequent w BIs.

It is worth noting that states whose backoff times are zero are referred to as *active states*, while the other states are referred to as *inactive states*.

In the above state transition diagram, the steady probability of state (r, w) is defined as

$$\pi_{r,w} = \lim_{t \to \infty} \mathbb{P}\left(C(t) = r, B(t) = w\right). \tag{4.6}$$

The corresponding steady-state probability of the proposed Markov chain is given by

$$\boldsymbol{\pi} = \{\pi_{0,0}, \pi_{1,0}, \ldots, \pi_{R-1,0}, \pi_{R,0}, \pi_{R,1}, \ldots, \pi_{R,W-1}\} \in \mathbb{R}^{(R+W) \times 1}.$$

Let $\mathbf{P} \in \mathbb{R}^{(R+W) \times (R+W)}$ denote the state transition matrix whose non-null elements are given in (4.1)–(4.5). The $\boldsymbol{\pi}$ can be obtained via mathematically solving the following balance equations:

$$\mathbf{P}\boldsymbol{\pi} = \boldsymbol{\pi} \tag{4.7a}$$

$$\sum_{r=0}^{R} \sum_{w=0}^{W-1} \pi_{r,w} = 1. \tag{4.7b}$$

Here, (4.7b) is due to the fact that the summation of all steady-state probabilities should equal to one.

> **Remark 4.3** It is worth noting that our model is *distinct* from the celebrated Bianchi's model [18] in the following two ways:
>
> - STAs backoff after every collision in the Bianchi's model, while STAs only backoff when the consecutive collision counter exceeds the retry limit in our proposed analytical model.
> - The contention window size increases with the number of consecutive collisions in the Bianchi's model, while the contention window size is fixed in our model.
>
> These distinctions are due to the different behaviors of two MAC protocols.

4.4.2 Performance Analysis

We first analyze the successful beamforming training probability in Sect. 4.4.2.1, based on which we analyze the beamforming training latency in Sect. 4.4.2.2 and the protocol throughput in Sect. 4.4.2.3. Particularly, we further analyze the maximum throughput via an asymptotic analysis method in Sect. 4.4.2.4.

4.4.2.1 Successful Beamforming Training Probability Analysis

To obtain the successful beamforming training probability, we first derive the closed-form expression of the steady-state probability and then obtain the successful beamforming training probability.

The steady-state probability is first derived via the following theorem.

Theorem 4.1 *The steady-state probabilities in the proposed Markov chain based model can be given by*

$$
\pi_{r,w} = \begin{cases} \dfrac{p^r (1 - p)}{p^R (W - 1)/2 + 1}, \forall r \in [0, R - 1], w = 0 \\[3mm] \dfrac{(W - w) p^R}{W \left(p^R (W - 1)/2 + 1 \right)}, \forall r = R, w \in [0, W - 1]. \end{cases} \tag{4.8}
$$

Proof Based on the proposed analytical model, the steady-state probabilities can be solved via the following three steps:

- Firstly, given the one-step transition probability in (4.1), we know that $\pi_{r+1,0} = p \cdot \pi_{r,0}, \forall r \in [0, R-2]$. Therefore, the steady-state probability at state $(R-1, 0)$ can be represented by

$$\pi_{R-1,0} = p^{R-1}\pi_{0,0}. \qquad (4.9)$$

- Secondly, given the one-step transition probability in the backoff states (4.3)–(4.5), the steady probability of backoff states can be given by

$$\pi_{R,w} = \frac{(W-w)\,p}{W}\left(\pi_{R,0} + \pi_{R-1,0}\right), \forall w \in [0, W-1]. \qquad (4.10)$$

- Thirdly, it can be obtained that $\pi_{R,0} = \frac{p}{1-p}\pi_{R-1,0}$ by taking $w = 0$ in (4.10). Hence, based on (4.9), (4.10) can be rewritten as

$$\begin{aligned} \pi_{R,w} &= \frac{(W-w)\,p}{W\,(1-p)}\pi_{R-1,0} \\ &= \frac{(W-w)\,p^R}{W\,(1-p)}\pi_{0,0}, \forall w \in [0, W-1]. \end{aligned} \qquad (4.11)$$

Hence, Theorem 4.1 is proved. □

In Theorem 4.1, all steady-state probabilities are represented in terms of p. Here, p is the unknown conditional collision probability. In the following, the value of p is obtained via the following steps.

- Firstly, taking all the active states into consideration, the probability that an STA stays in an active state is given by

$$\begin{aligned} \tau &= \sum_{r=0}^{R} \pi_{r,0} \\ &= \frac{1}{p^R\,(W-1)/2 + 1}. \end{aligned} \qquad (4.12)$$

- Secondly, for an STA, a successful transmission event occurs, only when the other active STAs select other A-BFT slots for transmission. Therefore, given an STA is active, the conditional successful transmission probability is represented by

$$\begin{aligned} p_s &= \left(\tau\left(1 - \frac{1}{M}\right) + 1 - \tau\right)^{N-1} \\ &= \left(1 - \frac{\tau}{M}\right)^{N-1} \\ &= \left(1 - \frac{1}{M\left(p^R\,(W-1)/2 + 1\right)}\right)^{N-1}. \end{aligned} \qquad (4.13)$$

- The last step is due to the substitution of (4.12). Since $p_s + p = 1$, the following equation can be obtained:

$$\left(1 - \frac{1}{M\left(p^R\left(W-1\right)/2+1\right)}\right)^{N-1} + p - 1 = 0. \tag{4.14}$$

The value of p can be obtained by solving (4.14).

We can see that (4.14) is an implicit function due to the summation and permutation. Thus, it is challenging to obtain a closed-form solution. To address this issue, we apply a numerical method to obtain p. More importantly, (4.14) shows that p depends on many protocol parameters, including retry limit R, contention window W, the number of A-BFT slots M, and the number of STAs N. Moreover, these protocol parameters are closely coupled with each other. Hence, the analysis on the network throughput is challenging.

With the obtained p, the successful transmission probability can be computed. Note that $1 - p$ is the conditional successful transmission probability given the STA is active. So the successful transmission probability is represented by

$$\hat{p}_s = (1-p)\,\tau. \tag{4.15}$$

The above value also denotes the probability of successful beamforming training in BFT-MAC.

4.4.2.2 Average Beamforming Training Latency Analysis

In addition to the average successful beamforming training probability, the beamforming training latency is also another important performance metric for an MAC protocol, especially for delay-sensitive applications. *The average beamforming training latency is defined as the average time taken until a successful transmission event occurs.*

The average beamforming training latency needs to take the consecutive collisions before a successful transmission into account, which can be represented by

$$D = \sum_{i=0}^{\infty} \mathbb{P}\left(\text{Success}|\text{Collisions} = i\right) \mathbb{E}\left[D_i\right]$$
$$= \sum_{i=0}^{\infty} (1-p)\, p^i \mathbb{E}\left[D_i\right]. \tag{4.16}$$

Here, $\mathbb{E}[D_i]$ represents the beamforming training latency of a successful transmission event after experiencing i consecutive collisions. It is worth noting that the STA would be frozen from transmission for a backoff time when the number of consecutive collisions exceeds the retry limit. As such, based on the number of consecutive collisions, $\mathbb{E}[D_i]$ can be represented by the following two cases:

- Case I: When $i < R$, $\mathbb{E}[D_i]$ can be represented by

$$
\begin{aligned}
\mathbb{E}[D_i] &= i \cdot T_{BI} + F \cdot T_{SSW} \\
&= T_{BI}\,(i + \alpha)
\end{aligned}
\tag{4.17}
$$

where $\alpha = F \cdot T_{SSW}/T_{BI}$. Here, T_{BI} and T_{SSW} denote the duration of a BI and an SSW frame, respectively. The above equation consists of the following terms: (1) The first term in (4.17) represents the latency caused by i consecutive collisions before a successful transmission event occurs. According to BFT-MAC, when a collision occurs, the collided STA must wait until the subsequent BI to initiate a transmission attempt. As such, the latency is increased by a BI. (2) The second term in (4.17) denotes the latency taken for the successful transmission. As illustrated in Fig. 4.2, the successful beamforming training procedure consists of F SSW frames.[2] The corresponding beamforming training latency is $F \cdot T_{SSW}$. With (4.17), when the number of consecutive collisions is less than R, the average beamforming training latency of an STA is represented by

$$
\begin{aligned}
\sum_{i=0}^{R-1} (1-p)\,p^i \mathbb{E}[D_i] &= \sum_{i=0}^{R-1} (1-p)\,p^i\,(i+\alpha)\,T_{BI} \\
&= T_{BI}\,(1-p)\left(\sum_{i=0}^{R-1} p^i \cdot i + \alpha \sum_{i=0}^{R-1} p^i \right) \\
&= T_{BI}\left(\frac{p^{R+1}\,(R-1) - Rp^R + p}{1-p} + \left(1 - p^R\right)\alpha \right).
\end{aligned}
\tag{4.18}
$$

- Case II: When $i \geq R$, each collision results in a further random backoff time of w BIs. In this case, $\mathbb{E}[D_i]$ is represented by

$$
\begin{aligned}
\mathbb{E}[D_i] &= ((i - R + 1)(\mathbb{E}[w] + 1) + R - 1)T_{BI} + F \cdot T_{SSW} \\
&= T_{BI}\,((i - R + 1)(\mathbb{E}[w] + 1) + R - 1 + \alpha).
\end{aligned}
\tag{4.19}
$$

Here, $\mathbb{E}[w]$ denotes the average backoff time. When the number of a consecutive collision exceeds R, the beamforming training latency can be represented by

---

[2] Here, F is referred to as *FSS* field in 802.11ad.

$$\sum_{i=R}^{\infty} (1-p) \, p^i \mathbb{E}\,[D_i]$$

$$= \sum_{i=R}^{\infty} T_{BI} \, (1-p) \, p^i \, ((i-R+1)(\mathbb{E}\,[w]+1) + R - 1 + \alpha)$$

$$\overset{(a)}{=} \sum_{j=0}^{\infty} T_{BI} \, (1-p) \, p^{R+j} \, ((j+1) \, (\mathbb{E}\,[w]+1) + R - 1 + \alpha)$$

$$= T_{BI} \, (1-p) \, p^R \left((\mathbb{E}\,[w]+1) \sum_{j=0}^{\infty} p^j \cdot j \right. \tag{4.20}$$

$$\left. + ((\mathbb{E}\,[w]+1) + R - 1 + \alpha) \sum_{j=0}^{\infty} p^j \right)$$

$$= T_{BI} \cdot p^R \left(\frac{p(\mathbb{E}\,[w]+1)}{1-p} + \mathbb{E}\,[w] + R + \alpha \right)$$

$$\overset{(b)}{=} T_{BI} \cdot p^R \left(\frac{W+1}{2(1-p)} + R + \alpha - 1 \right).$$

The derivations in the above equation are due to the following fact: (a) follows by changing variable $j = i - R$; and (b) is due to the substitution of $\mathbb{E}\,[w] = (W-1)/2$.

Overall, taking these two cases into consideration, the average beamforming training latency in (4.16) can be obtained as follows:

$$D = \sum_{i=0}^{R-1} (1-p) \, p^i \mathbb{E}\,[D_i] + \sum_{i=R}^{\infty} (1-p) \, p^i \mathbb{E}\,[D_i]$$

$$= T_{BI} \left(\frac{p^{R+1} \, (R-1) - Rp^R + p}{1-p} + \left(1 - p^R\right) \alpha \right)$$

$$+ T_{BI} p^R \left(\frac{W+1}{2(1-p)} + R + \alpha - 1 \right) \tag{4.21}$$

$$= T_{BI} \left(\frac{p^R (W-1)/2 + p}{1-p} + a \right).$$

The above equation shows that collision probability p, retry limit R, and the contention window W impact the beamforming training latency. Obviously, we can see that latency increases with the collision probability. This is because severe collisions would result in substantial retransmission in the network and increase the latency.

4.4.2.3 Normalized Throughput Analysis

In the following, we analyze the throughput of BFT-MAC. Based on the previous analysis, the average successful transmission probability is $p_s \tau$, and hence the average number of STAs that successfully perform beamforming training is represented by $p_s \tau N$. *The normalized throughput of BFT-MAC is defined as the percentage of A-BFT slots that has been successfully utilized.*

The throughput reflects the beamforming training capability of the MAC protocol, which is given by

$$
\begin{aligned}
S &= \frac{p_s \tau N}{M} \\
&= \left(1 - \frac{\tau}{M}\right)^{N-1} \frac{\tau N}{M}
\end{aligned}
\tag{4.22}
$$

where the last step follows from the substitution of (4.13). Note that τ depends on protocol parameters. Hence, the above equation characterizes the impact of the number of STAs, the number of A-BFT slots, and MAC parameters on the MAC throughput.

From a simple analysis on (4.22), we can see that obtaining the asymptotic MAC throughput in dense user scenarios is computational complex. The reason is that (4.22) is a complicate function of τ, N, and M. To address the issue, an approximation of throughput is desired to acquire tractable performance analysis. With results in (4.13), when the number of STAs is large, the conditional successful transmission probability can be approximated by

$$
\begin{aligned}
\hat{p}_s &= \left(1 - \frac{\tau}{M}\right)^{N-1} \\
&\overset{(a)}{\approx} \left(1 - \frac{\tau}{M}\right)^{N} \\
&= \left(\left(1 - \frac{\tau}{M}\right)^{M/\tau}\right)^{N\tau/M} \\
&\overset{(b)}{\approx} e^{-N\tau/M}.
\end{aligned}
\tag{4.23}
$$

In the above equation, (a) is obtained when N is sufficiently large. In this chapter, we consider dense user scenarios, and the condition can be easily satisfied. Hence, the approximation is reasonable. Here, (b) follows from the equation

$$\lim_{n \to \infty} \left(1 - \frac{1}{n} \right)^n = \frac{1}{e}$$

where $n = M/\tau$ is sufficiently large in dense user scenarios. From (4.23), we can find that the average successful transmission probability is dependent on N/M.

Substituting (4.23) into (4.22), when the number of STAs is sufficiently large, the *asymptotic normalized throughput* can be represented by

$$\hat{S} = \frac{\tau N}{M} e^{-\tau N/M}. \tag{4.24}$$

The above equation characterizes the asymptotic throughput performance with respect to system parameters. In addition, the simulation results in Fig. 4.8 have validated the accuracy of this approximation.

Remark 4.4 Based on the above analysis, we show the following important insights for MAC design in dense user scenarios:

- Firstly, the ratio between the number of STAs and the number of A-BFT slots, i.e., N/M, impacts the normalized throughput in dense user scenarios. As such, it is an effective solution to increase the number of A-BFT slots adaptive to the number of STAs, thereby maintaining excellent performance.
- Secondly, given a further analysis of (4.24), when $\tau N/M$ is less than 1, the throughput increases with $\tau N/M$. When $\tau N/M$ exceeds 1, the throughput decreases with $\tau N/M$. In practical systems with a fixed number of A-BFT slots, we find that the throughput would decrease with the increase of user density in dense user scenarios.
- Thirdly, as the MAC parameters determine τ, the MAC parameters also affected the throughput. Thus, we claim that the default MAC parameter setting may become suboptimal when user density changes over time. The statement implies that the MAC parameters should be tuned according to the user density.

4.4.2.4   Maximum Normalized Throughput Analysis

Given the aforementioned asymptotic normalized throughput analysis, we aim to analyze the *maximum normalized throughput* in dense user scenarios. The number of A-BFT slots can be optimized with respect to the number of STAs to maximize the normalized throughput, since the number of STAs in the network is uncontrollable. For simplicity, the number of A-BFT slots is assumed to be sufficient. The case with limited number of of A-BFT slots is discussed in Sect. 4.4.2.5.

In this case, we formulate the normalized throughput maximization problem as follows:

$$\mathcal{P}2 : \max_{M} \quad \hat{S}$$

$$\text{s.t.} \quad M \in \mathbb{Z}^{+}. \tag{4.25}$$

Note that the number of A-BFT slots should take positive integers as indicated by the constraint. Obviously, problem (4.25) is an integer programming problem. To address the above problem, the integer constraint is first relaxed to a non-integer constraint, and then this optimization problem can be solved by taking the derivation of (4.24). The optimal value of A-BFT slots for achieving the maximum normalized throughput is denoted by

$$M^{\star} = \tau N. \tag{4.26}$$

The above Eq. (4.26) provides an interesting insight on the MAC design. For maximizing the normalized throughput, the number of A-BFT slots should equal the number of active STAs (τN) in the network. In other words, the network needs to provide equivalent A-BFT slots for all active users.

Given the optimal condition in (4.26), the maximum normalized throughput is represented by

$$\hat{S}^{\star} = e^{-1} \tag{4.27}$$

The value of the throughput is the same as that of slotted ALOHA.

The optimal number of A-BFT slots is dependent on the MAC parameters since τ is related to the MAC parameters. Once the condition in (4.26) is satisfied, we can have $p = 1 - 1/e$ because of (4.23). Then, based on (4.12), (4.26) can be rewritten as

$$M^{\star} = \frac{N}{\left(1 - e^{-1}\right)^{R} (W - 1) /2 + 1}. \tag{4.28}$$

The above equation characterizes the relationship between MAC parameters R and W and the optimal number of A-BFT slots. By analyzing (4.28), we find that the optimal number of A-BFT slots decreases with the decrease of R and the increase of W. Hence, a small value of R and a large value of W should be chosen to achieve the maximum throughput for a limited number of A-BFT slots. In Sect. 4.4.2.5, the optimization of MAC parameters is given.

Remark 4.5 The above asymptotic throughput analysis illustrates several useful insights onto the performance of BFT-MAC:

- Firstly, the maximum normalized throughput achieved by BFT-MAC is barely $1/e$. The value is the same as that of slotted ALOHA. The low normalized throughput is due to the absence of carrier sensing mechanism. The lack of carrier sensing mechanism leads to severe collisions in beamforming training.
- Secondly, the optimal number of A-BFT slots for achieving the maximum throughput should equal the number of active STAs in the network. In other words, the mismatch between the active STAs and A-BFT slots results in the throughput degradation.
- Finally, analytical results indicate that the protocol parameters R and W also impact the normalized throughput.

4.4.2.5   Enhancement Scheme for 802.11ad BFT-MAC

In the following, to improve MAC performance in dense user scenarios, we propose an enhancement scheme. The previous analysis results indicate that the ratio between the number of STAs and the number of A-BFT slots impacts the normalized throughput. However, in practical mmWave WLANs, the number of A-BFT slots is limited. For instance, in the current 802.11ad standard, at most, eight A-BFT slots are provided for beamforming training. In the future 802.11ay standard, the number of A-BFT slots is expected to increase up to 40 [30]. The number of A-BFT slots is still limited, as compared to the time-varying number of STAs. In this way, due to the limitation of A-BFT slots, the throughput degradation can be observed in dense user scenarios.

In this chapter, we propose an *enhancement scheme* to improve the throughput in dense user networks. In the enhancement scheme, MAC parameters R and W should be tuned with the user density in the network. The reason is that tuning MAC parameters can determine the probability that STAs stay active (τ). As such, the number of active STAs could be guaranteed to be equivalent to the number of the provided A-BFT slots. Our analytical results also validate the reason. As shown in (4.12), we have

$$\tau = \frac{1}{\left(p^R \left(W - 1\right)/2 + 1\right)}.$$

Here, with the decrease of R, the value of τ decreases. The fact means that a small value of the retry limit results in that STAs are prone to enter backoff states, such that reducing the number of active STAs in the network. In addition, with the decrease of W, the value of τ also decreases. This is because a large value of the contention window size renders STAs enter inactive states for a long time. Therefore, it is an effective solution to adaptively adjust MAC parameters in tune with user density to provide satisfactory performance in dense user scenarios.

The *operation* of the proposed enhancement scheme consists of the following two steps:

- Firstly, the AP collects the information on the number of STAs in the network. The information can be easily obtained in commercial WLANs.
- Secondly, according to the collected information on the number of STAs in the network, the AP determines the optimal BFT-MAC parameter configuration based on the optimization algorithm and then broadcasts the optimal MAC parameter configuration, including the retry limit and the contention window, to all the STAs in the network.

The key issue in the proposed enhancement scheme is the optimal MAC parameter configuration. The issue is formulated as the following optimization problem with the objective of maximizing the normalized throughput:

$$\mathcal{P}3 : \max_{R,W} \quad S$$

$$\text{s.t.} \quad 1 \leq W \leq W_{max}, W \in \mathbb{Z}^+ \tag{4.29a}$$

$$1 \leq R \leq R_{max}, R \in \mathbb{Z}^+. \tag{4.29b}$$

Here, W_{max} represents the maximum contention window size. This value is adopted to avoid infinite backoff time which may result in significant beamforming training latency. Here, R_{max} represents the maximum value of the retry limit.

The above optimization problem is *challenging* to be solved due to the following two reasons:

- Firstly, the optimization variables need to follow the integer constraints, and hence the problem is an integer optimization problem.
- Secondly, the objective function is non-convex. The reason is that S is an implicit function in terms of two coupled variables R and W. According to (4.14), it can be clearly seen that variables R and W are coupled with each other.

Overall, the above problem is an integer non-convex optimization problem, and obtaining the optimal solution is difficult via traditional optimization methods.

We solve the problem by leveraging an inherent property of the problem. We find that the number of possible combinations of the MAC parameters is limited. As such, an *exhaustive search method* can be applied to solve this problem [19]. The computational complexity of the exhaustive search method can be easily analyzed, i.e., $O(W_{max} R_{max})$. Moreover, the optimal parameter setting can be computed with different numbers of STAs in an offline manner, and the computed parameter setting is loaded into AP as a table. As such, the computational time can be further reduced. In the online operation, according to the number of STAs collected by the AP, the AP could search the table to determine the optimal MAC parameter setting with a low complexity.

4.5 Performance Evaluation for 802.11ad BFT-MAC

In this section, we evaluate the proposed analytical model and the enhancement scheme via extensive Monte-Carlo simulations. We first provide the simulation setup in Sect. 4.5.1 and then validate the effectiveness of the proposed analytical model under different parameter settings in terms of different performance metrics in Sect. 4.5.2. Finally, we demonstrate the performance gain of the proposed enhancement scheme in dense network scenarios in Sect. 4.5.3.

4.5.1 Simulation Setup

The proposed analytical model is validated via a discrete event simulator coded in MATLAB. An 802.11ad system which operates at the unlicensed 60 GHz frequency band is considered in the simulation. Specifically, 10,000 BIs are simulated, and the observed statistics of interests are studied. Unless otherwise specified, based on the default parameter configuration of the 802.11ad standard [32], we set $M = 8$, $W = 8$ and $R = 8$. Important simulation parameters are given in Table 4.2. In each experiment, 1000 simulation runs are conducted, and a 95% confidence interval for each simulation point is plotted.

Table 4.2 Simulation parameters in the BFT-MAC protocol

Parameter	Value
BI duration (T_{BI})	100 ms
SSW frame duration (T_{SSW})	15.8 us
Number of STAs (N)	[8, 32]
Number of A-BFT slots (M)	8
Default retry limit (R)	8
Default contention window (W)	8
Number of SSW frames in an A-BFT slot (F)	16
Frequency band	60 GHz
Maximum retry limit (R_{max})	10
Maximum contention window (W_{max})	10

4.5.2   Validation of Analytical Model

As shown in Fig. 4.4, we show the successful beamforming training probability
with respect to the number of STAs when $M = 8, 12$, and 16. It is obvious that
the results obtained via our analytical model are highly consistent with that via
simulations, thereby validating the accuracy of our analytical model. As expected, it
can be seen that a lower successful beamforming training probability is observed in
denser user scenarios. This is because more STAs contend for limited A-BFT slots.
Specifically, when the number of STAs increases from 4 to 32, it can be clearly
seen that the successful beamforming training probability drops from more than
80% to less than 20%. Moreover, as the number of A-BFT slots, M, increases,
the successful beamforming training probability increases since more A-BFT slots
are provided to beamforming training. The above results validate our theoretical
result that the successful beamforming training probability mainly depends on the
numbers of STAs and A-BFT slots.

As shown in Fig. 4.5, the normalized throughput performance with respect to
the number of STAs is presented. A few important observations can be made.
Firstly, simulation results are closely matched with analytical results, which further
validates our analytical model. Secondly, the normalized throughput exhibits a bell-
shaped behavior, which is due to the following two reasons: (1) Many A-BFT slots
are not utilized in low user density scenarios, and (2) severe collision occurs in
high user density scenarios, which leads to a low throughput. Therefore, in order
to achieve the maximum normalized throughput, we should cautiously select the
number of A-BFT slots. Thirdly, we can see that a system with more A-BFT slots

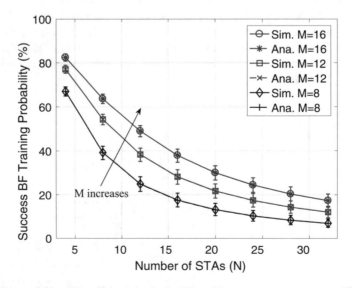

Fig. 4.4 Successful beamforming training probability with respect to the number of STAs

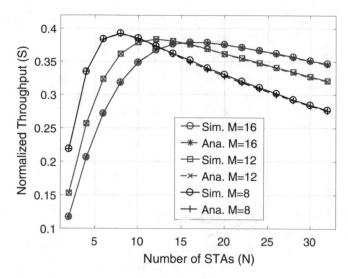

Fig. 4.5 Normalized throughput with respect to the number of STAs

achieves a higher normalized throughput than that with fewer A-BFT slots. For instance, in a dense user scenario, i.e., $N = 32$, the normalized throughput for $M = 16$ is about 25%, which is more than that for $M = 8$. The reason is that more A-BFT slots can effectively alleviate the collision issue in dense user scenarios. Finally, we can observe the maximum normalize throughput is around $1/e$ (i.e., about 0.37), which also complies with our analytical results.

As shown in Fig. 4.6, the impact of the number of STAs on the average beamforming training latency is evaluated. It can be observed that simulation results comply with our analytical results in (4.21). It is clear that the beamforming training latency increases with the number of STAs. The reason is that STAs suffer from severe collisions in dense user scenarios, resulting in a large amount of retransmission. Furthermore, as the number of A-BFT slots increases, the average beamforming training latency decreases. When N is 32, the average beamforming training latency for $M = 8$ is up to 1.3 s, which is 150% more than that for $M = 16$. Therefore, increasing the number of A-BFT slots can reduce the beamforming training latency.

As shown in Fig. 4.7, we plot the impact of the value of the retry limit on the normalized throughput. We first observe that the normalized throughput of BFT-MAC varies with different values of retry limit. Specifically, BFT-MAC with a small value of the retry limit achieves a higher throughput than that with a larger value of retry limit in dense user scenarios. For instance, when the number of users is 32, BFT-MAC for $R = 2$ achieves around 28% throughput gain as compared to that for $R = 8$. This is because a small value of the retry limit leads to STAs which are susceptible to enter the backoff states. In this way, fewer active STAs contend for the A-BFT slots, hence enhancing the performance in dense user scenarios. As

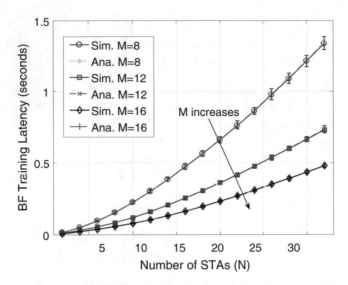

Fig. 4.6 Average beamforming training latency with respect to different numbers of STAs

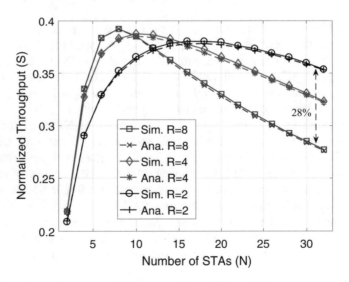

Fig. 4.7 Normalized throughput with different values of the retry limit

$R = 8$ is the default configuration of the retry limit, we know that the default MAC configuration is suboptimal in dense user scenarios. Hence, to improve MAC performance, adaptively adjusting MAC parameters in tune with the user density is a potential solution.

As shown in Fig. 4.8, we study the normalized throughput in terms of the ratio between the number of STAs and the number of A-BFT slots. We can see that the normalized throughput with the same ratio for different numbers of A-BFTs and

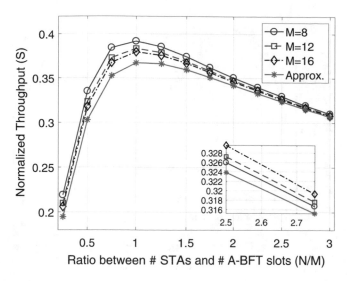

Fig. 4.8 Normalized throughput with respect to different ratios between the number of STAs and the number of A-BFT slots

STAs is quite close. The simulation results validate our asymptotic analysis results, i.e., the MAC throughput mainly depends on the ratio between contending STAs and the provided A-BFT slots. In addition, with the increase of M, the gap between simulation results and our approximation in (4.24) narrows. The gap is negligible in dense user scenarios, i.e., when the ratio is larger than 2, which indicates the accuracy of our approximation.

4.5.3   Enhancement Scheme Evaluation

In this subsection, performance of the proposed enhancement scheme is evaluated, by comparing with two benchmarks:

- Default 802.11ad BFT-MAC protocol, in which both R and W adopt default parameter configurations
- Slotted ALOHA protocol, in which each STA randomly selects an A-BFT slot for beamforming training without a backoff mechanism

As shown in Figs. 4.9 and 4.10, we compare the normalized throughput of the proposed enhancement scheme with the benchmarks. Several important observations can be obtained from simulation results:

- Firstly, the proposed enhancement scheme can significantly enhance the normalized throughput in dense user scenarios, compared with the benchmarks. The simulation results are presented in Fig. 4.9. Specifically, a 35% performance gain

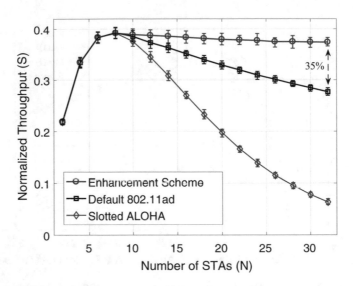

Fig. 4.9 Normalized throughput comparison for $M = 8$

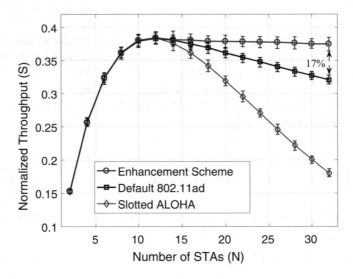

Fig. 4.10 Normalized throughput comparison for $M = 12$

can be achieved when the number of users is 32. The observation demonstrates the effectiveness of the proposed enhancement scheme. The underlying reason is that the MAC parameters in the proposed scheme can be adjusted in tune with user density to achieve the best performance.

- Secondly, nearly the same performance is achieved by all three schemes in low user density scenarios. The key difference between three schemes is the backoff mechanism, which mainly works in dense user scenarios and impacts the performance there.

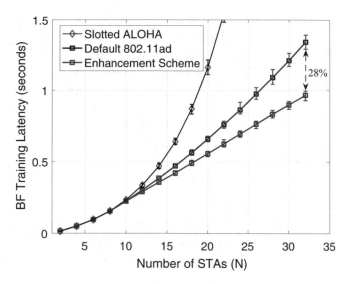

Fig. 4.11 Average beamforming training latency comparison for $M = 8$

- Thirdly, as the number of A-BFT slots increases, the performance gain decreases. As shown in Fig. 4.10, when the number of A-BFT slots is 12, the proposed enhancement scheme only achieves 17% performance gain, as compared to the default 802.11ad. Therefore, it is more suitable to apply the proposed enhancement scheme in dense user scenarios.

Next, the average beamforming training latency comparison between the proposed scheme and benchmarks for $M = 8$, and 12 is shown in Figs. 4.11 and 4.12, respectively. We can see that the average beamforming training latency of all three schemes increases with the number of STAs, since the collision probability increases. In addition, the slotted ALOHA protocol suffers from severe latency as compared to benchmarks. The slotted ALOHA lacks a backoff mechanism, and hence all STAs always stay active, which results in a higher collision probability and a longer delay. Compared with the default 802.11ad, the proposed scheme can achieve a considerable latency reduction. Specifically, for the number of STAs is 32, a 28% performance gain can be observed clearly in Fig. 4.11. Similar to the normalized throughput, with the growth of the number of A-BFT slots, the performance gain on beamforming training latency reduces as the increase of A-BFT slots relieves the collision. As plotted in Fig. 4.12, for $M = 12$, the proposed scheme can still obtain a 16% latency reduction when the number of STAs is 32. The result further validates the effectiveness of the proposed scheme. Even though the beamforming training latency can be reduced via adjusting MAC parameters, hundreds of milliseconds latency are still incurred in dense user scenarios.

Figure 4.13 plots the optimal values of the retry limit in terms of the number of STAs when $M = 8, 12$, and 16. Clearly, the optimal value of the retry limit decreases with the number of STAs. The results show that a small value of retry

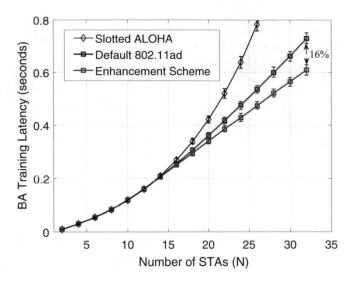

Fig. 4.12 Average beamforming training latency comparison for $M = 12$

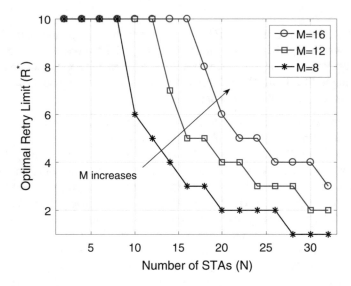

Fig. 4.13 The optimal value of the retry limit in terms of the number of STAs

limit is preferred in dense user scenarios. For instance, for the number of A-BFT slots is 8, the optimal value of the retry limit decreases to 1, when the number of STAs is larger than 28. This is because a small value of the retry limit in dense user scenarios leads to those STAs prone to enter backoff states. As such, the collision probability is reduced, thereby enhancing the normalized throughput. Moreover, the network becomes less congested with the increase of A-BFT slots. Hence, we should chose a large value of the retry limit. For example, when the number of STAs is 32, the optimal retry limit value is 3 for the number of A-BFT slots is 16. The value is larger than that for the number of A-BFT slots is 8.

4.6 Multiuser Beamforming Training Protocol Design and Analysis

The previous sections belong to the first part of this chapter, which focuses on analyzing and enhancing beamforming training performance in the current 802.11ad standard, i.e., the single-user transmission scenario. The following two sections are the second part of the chapter, which focuses on beamforming training protocol design and performance analysis in the multiuser transmission scenario.

With large communication bandwidth, mmWave communication can achieve a transmission data rate up to multiple gigabits per second, which facilitates mmWave Wi-Fi systems to support data-hungry applications, such as wireless docking, high-definition video transmission, and real-time virtual reality gaming. *Multiuser transmission is a technology to transmit data to multiple STAs simultaneously.* By exploiting the multiuser multiplexing gain, the aggregated data rate increases approximate linearly with the number of users that perform multiuser transmission.

The support of multiuser transmission is an important feature for high-speed wireless networks. For microwave WLANs, the current standard, i.e., 802.11ac, can support up to four users. Such multiuser transmission feature is expected to be inherited in the next-generation WLAN standard, i.e., 802.11ax. For mmWave WLANs, however, current 802.11ad can only support single-user transmission. Toward future mmWave WLANs, the next-generation standard, i.e., IEEE 802.11ay, considers multiuser transmission as the key technology to improve throughput. Current 802.11ad achieves a peak data rate approximately 7 Gbps, while 802.11ay is going to witness a sky-rocketing throughput increase. The peak data rate in 802.11ay is expected to up to 40 Gbps [30, 31, 34].

In mmWave communication, a large antenna array is utilized to obtain sufficient directional antenna gain to conquer extremely high path loss. However, the large antenna array poses new challenges to support multiuser transmission via conventional digital beamforming methods. The conventional digital beamforming methods require real-time channel state information between the transmitter and the receiver. Operating such methods in mmWave systems would incur high overhead in channel estimation and excessive power consumption with fully digital precoding since the channel state information is huge for a large antenna array with a number of antenna elements. Thus, an efficient multiuser beamforming scheme is desired for mmWave communication.

In the following, we propose a multiuser beamforming training protocol and analyze its performance in Sect. 4.6, and then simulation results are provided to validate the performance of the proposed protocol in Sect. 4.7. Specifically, in this section, we first give a review on the existing multiuser transmission schemes in Sect. 4.6.1. Then, we design a multiuser transmission scheme for downlink mmWave communication in Sect. 4.6.2. Then, we propose a tailored multiuser beamforming training protocol to support multiuser transmission in Sect. 4.6.3 and analyze the overhead of the proposed protocol in Sect. 4.6.4.

4.6.1 Existing Works on Multiuser Transmission

Existing works propose multiple schemes to enable multiuser transmission in mmWave communication. Among them, hybrid beamforming is considered as a promising solution with a low complexity [14, 35, 36]. Hybrid beamforming consists of two parts, i.e., analog beamforming and digital beamforming. The analog beamforming part controls the phase of the transmitted signal via analog phase shifters to provide high antenna gain for addressing path loss issues. The digital beamforming part designs the baseband beamforming matrix to mitigate multiuser inference among users. Existing theoretical analysis and empirical experiment results have shown that hybrid beamforming can achieve close-to-optimal performance as compared to fully digital beamforming benchmark with a much lower complexity [15]. For more details on hybrid beamforming, one can refer to a detailed introduction in Sect. 2.2.3 in Chap. 2 in this monograph.

Many works have been devoted to designing and analyzing hybrid beamforming in different mmWave settings. Alkhateeb et al. in [37] proposed a hybrid precoding in wideband mmWave communication systems with a feedback between transmitter and receiver. The sparsity of received signal is exploited to design an efficient hybrid precoding scheme in [38]. This method is extended work to mmWave cellular networks [39]. In addition to traditional cellular networks, hybrid beamforming schemes are also developed for high-mobility communications scenarios. A novel hybrid beamforming scheme is developed for the high-speed railway communication [40]. The proposed scheme leverages a number of train-mounted mobile relay to enable multiuser transmissions, thereby providing high-speed downlink communication services for the train. In another direction, the user selection issue is investigated. In the hybrid beamforming, multiple users should be selected to offer multiuser transmission simultaneously. A low complexity quasi-orthogonal user selection among a number of users based on their beam index information is proposed in [41, 42], and the corresponding performance analysis is provided therein.

4.6.2 Multiuser Transmission Scheme

In the following, we first show hybrid beamforming architecture in Sect. 4.6.2.1 and then present the corresponding hybrid beamforming algorithm in Sect. 4.6.2.2.

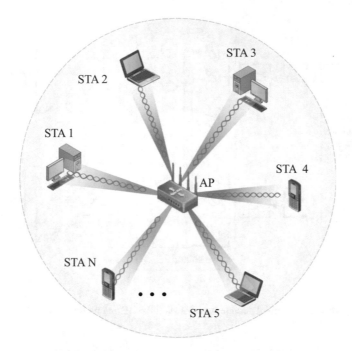

Fig. 4.14 Considered scenario for multiuser transmission, in which several STAs transmit data to the AP simultaneously

4.6.2.1 Hybrid Beamforming Architecture

As shown in Fig. 4.14, we consider a downlink multiuser transmission scenario. To this end, we adopt the following hybrid beamforming architecture [16], as shown in Fig. 4.15. In addition to the analog beamforming at both the transmitter and the receiver, AP has digital beamforming whose function is to mitigate multiuser interference among users. Let N_s, N_{RF}, N_{BS}, and N_{MS} denote the number of transmitted data streams, the number of radio-frequency (RF) chains, the number of antennas at the AP, and the number of antennas at each user, respectively. For simplicity of analysis, users are assumed to be equipped with the same number of antennas. Let U denote the number of simultaneously supported users. For U users, we need to have U RF chains at the AP to support U data streams from the AP to the users. Hence, we have $N_s = N_{RF} = U$ [16].

Digital beamforming at the AP can be represented by the following matrix, i.e.,

$$\mathbf{F}_{BB} = [\mathbf{f}_1^{BB}, \mathbf{f}_2^{BB}, \dots, \mathbf{f}_U^{BB}] \in \mathbb{C}^{U \times U},$$

where each element represent the digital beamforming vector for a user. The analog beamforming at the AP is represented by the following matrix

$$\mathbf{F}_{RF} = [\mathbf{f}_1^{RF}, \mathbf{f}_2^{RF}, \dots, \mathbf{f}_U^{RF}] \in \mathbb{C}^{N_{BS} \times U}.$$

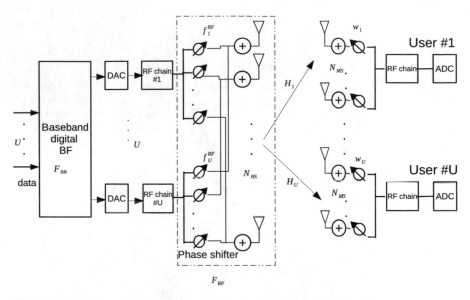

Fig. 4.15 Hybrid beamforming architecture between the AP and multiple users

As such, the transmitted signal after analog beamforming and digital beamforming at the AP is given by

$$\mathbf{x} = \mathbf{F}_{RF}\mathbf{F}_{BB}\mathbf{s}, \qquad (4.30)$$

where $\mathbf{s} = [s_1, s_2, \ldots, s_U]^T \in \mathbb{C}^{U \times 1}$ is the transmitted symbols for different users. The transmission power at the AP is given by $\mathbb{E}[|s_i|^2] = P/U$. Note that in this section, $\mathbf{A}^H$ and $\mathbf{A}^T$ stand for conjugate transpose and transpose of matrix $\mathbf{A}$, respectively. $\mathbf{a}^*$ represents the conjugate of vector $\mathbf{a}$.

The channel matrix between the AP and user i is represented by matrix $\mathbf{H}_i \in \mathbb{C}^{N_{MS} \times N_{BS}}$. Here, we consider a line-of-sight (LOS) path scenario, in which there is only one path between the AP and the user. In this way, the corresponding channel matrix $\mathbf{H}_i$ is given by Alkhateeb et al. [14]

$$\mathbf{H}_i = \sqrt{N_{MS}N_{BS}}a\boldsymbol{\alpha}_i(\theta)\boldsymbol{\alpha}_{BS}^*(\phi), \qquad (4.31)$$

where a represents the complex channel gain of the path. Here, $\boldsymbol{\alpha}_i(\theta)$ and $\boldsymbol{\alpha}_{BS}(\phi)$ are the antenna array response vectors of user i and the AP, respectively. Considering uniform antenna arrays, they can be represented by

$$\boldsymbol{\alpha}_i(\theta) = \frac{1}{\sqrt{N_{MS}}}[1, e^{j\frac{2\pi}{\lambda}r\sin(\theta)}, \ldots, e^{j(N_{MS}-1)\frac{2\pi}{\lambda}r\sin(\theta)}]^T, \qquad (4.32)$$

and

$$\boldsymbol{\alpha}_{BS}(\phi) = \frac{1}{\sqrt{N_{BS}}}[1, e^{j\frac{2\pi}{\lambda}r\sin(\phi)}, \ldots, e^{j(N_{BS}-1)\frac{2\pi}{\lambda}r\sin(\phi)}]^T, \qquad (4.33)$$

respectively. Here, λ is the signal wavelength, and r is the distance between antenna elements.

Let matrix $\mathbf{W} = [\mathbf{w}_1, \mathbf{w}_2, \ldots, \mathbf{w}_U]$ represent the RF combining matrix at users, i.e., analog beamforming at users. At the uth user, the RF combiner $\mathbf{w}_u \in \mathbb{C}^{N_{MS} \times 1}$ is used to process the received signal, and thus the received signal is given by

$$y_u = \mathbf{w}_u^* \mathbf{H}_u \sum_{n=1}^{U} \mathbf{F}_{RF} \mathbf{f}_n^{BB} s_n + \mathbf{w}_n^* \mathbf{n}_u, \qquad (4.34)$$

where $\mathbf{n}_u$ is the complex Gaussian noise with variance N_o.

With the above results, the sum rate for multiple users can be given by

$$R_{sum} = \sum_{u=1}^{U} \log_2 \left(1 + \frac{\frac{P}{U}|\mathbf{w}_u^* \mathbf{H}_u \mathbf{F}_{RF} \mathbf{f}_u^{BB}|^2}{\frac{P}{U}\sum_{n \neq u}|\mathbf{w}_u^* \mathbf{H}_u \mathbf{F}_{RF} \mathbf{f}_n^{BB}|^2) + N_o} \right). \qquad (4.35)$$

In the downlink multiuser transmission scenario, the goal is to design digital beamforming, i.e., $\mathbf{F}_{BB}$, and analog beamforming, i.e., $\mathbf{F}_{RF}$ and $\mathbf{W}$, with the objective of maximizing the sum rate. Thus, we can formulate the following optimization problem:

$$\mathcal{P}4: \max_{\mathbf{F}_{RF}, \mathbf{F}^{BB}, \mathbf{W}} R_{sum} \qquad (4.36a)$$

In the above problem, analog beamforming $\mathbf{F}_{RF}$ and $\mathbf{W}$ should be implemented with beamsteering codebooks in practical mmWave systems [43], while digital beamforming $\mathbf{F}_{BB}$ can be fully digitalized. As such, this problem is a non-convex optimization problem, such that the closed form of the optimal solution is difficult to obtain.

4.6.2.2   Proposed Hybrid Beamforming Algorithm

In the following, a low-complexity hybrid beamforming algorithm is proposed to solve the problem. The proposed algorithm consists of two stages, an analog beamforming stage and a digital beamforming stage, which are detailed as follows:

- *Analog beamforming stage*: Each user and the AP select their analog beamforming, i.e., $\mathbf{f}_u^{RF}$ and $\mathbf{w}_u$, to maximize directional antenna gain for their connection. As such, the optimal analog beamforming can be obtained via

$$\left\{ \mathbf{f}_u^{RF}, \mathbf{w}_u \right\} = \arg \max \| \mathbf{w}_u^* \mathbf{H}_u \mathbf{f}_u^{RF} \|.$$

In this stage, each user and the AP conduct operation based on the beamforming training protocol in 802.11ad.

- *Digital beamforming stage*: At the beginning of the second stage, each user and its corresponding RF chain select the optimal analog beamforming to transmit data. Next, each user estimates its effective channel, i.e., $\bar{\mathbf{h}}_i = \mathbf{w}_i^* \mathbf{H}_i \mathbf{F}_{RF}$, then feeds the effective channel back to the AP. Based on the effective channel, the AP can calculate the digital beamforming based on a zero-forcing precoding algorithm. In this way, the optimal digital beamforming is given by

$$\mathbf{F}_{BB} = \bar{\mathbf{H}}^* \left(\bar{\mathbf{H}} \bar{\mathbf{H}}^* \right)^{-1}, \tag{4.37}$$

where

$$\bar{\mathbf{H}} = [\bar{\mathbf{h}}_1^T, \bar{\mathbf{h}}_2^T, \ldots, \bar{\mathbf{h}}_U^T]^T. \tag{4.38}$$

In this stage, each user feeds back the effective channel to the AP, and then the AP calculates digital beamforming based on the effective channel.

4.6.3 Multiuser Beamforming Training Protocol

In the following, we propose a multiuser hybrid beamforming protocol that is compatible with the 802.11ad standard. The proposed protocol consists of three stages, i.e., multiuser sector level sweep (SLS), multiuser ATI, and multiuser beam refinement protocol (BRP) stages. As discussed above, the designed hybrid beamforming algorithm includes an analog beamforming stage and a digital beamforming stage. For the analog beamforming stage, multiuser SLS and multiuser BRP stages are performed for STAs and their corresponding RF chains, which is to obtain the analog beamforming. For the digital beamforming stage, the effective channel is estimated and fed back to the AP during the multiuser BRP stage to acquire the digital beamforming.

The detailed operations in three stages are given as follows:

- *Multiuser SLS stage*: As shown in Fig. 4.16, SLS for each STA is performed individually at each A-BFT slot for interference avoidance consideration. Multiple frames, such as SSW, sector sweep feedback, and sector sweep ACK frames, are exchanged between the AP and STAs. The maximum number of the A-BFT slots is eight in 802.11ad.
- *Multiuser ATI stage*: The stage is to announce channel access allocation. As shown in Fig. 4.17, N STAs in the coverage of the AP are polled. Then, each STA responds the AP with a service period request (SPR) frame which contains

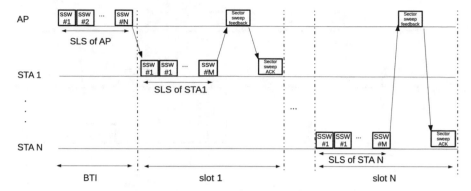

Fig. 4.16 An illustrative example of the multiuser SLS stage

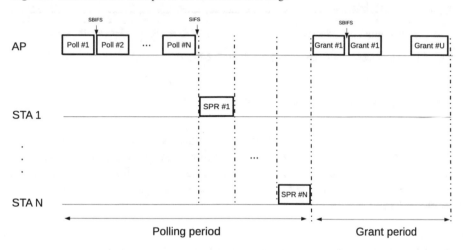

Fig. 4.17 An illustrative example of the multiuser ATI stage

information whether the STA requires to transmit data. Next, the AP grants U users to data transmission. In other words, N STAs are polled, and only U STAs are selected to transmit data simultaneously.

- *Multiuser BRP stage*: In this stage, as suggested in 802.11ad, each user needs to perform BRP with the AP to obtain the refined analog beamforming. As shown in Fig. 4.18, to avoid interference, BRP for different users are performed at different time slots. In addition, each user can estimate the channel when each user communicates to the AP via BRP frames. Then, at the end of multiuser BRP stage, each user can feed back the effective channel to the AP via BRP feedback frames. Based on the collected effective channel, digital beamforming can be solved according to (4.37).

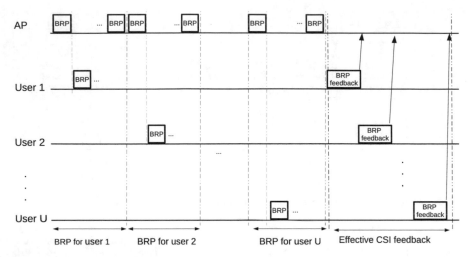

Fig. 4.18 An illustrative example of the multiuser BRP stage

4.6.4   Protocol Overhead Analysis

In mmWave communication, beamforming training incurs the non-negligible overhead for data transmission. Hence, we analyze the overhead of the proposed multiuser beamforming training protocol in terms of beamforming training time, as detailed in Sect. 4.6.4.2. For comparison, we first analyze the protocol overhead in the 802.11ad standard in Sect. 4.6.4.1.

4.6.4.1   Beamforming Training Time Analysis of 802.11ad Protocol

Aforementioned, in 802.11ad, the beamforming training process for one user consists of three parts: SLS, ATI, and BRP. The corresponding analysis is given as follows. Note that all the frames used in beamforming training are control frames. The control frames are transmitted with the control data rate, i.e., 27.5 Mbps, as suggested by 802.11ad.

- *SLS phase for one user*: The SLS phase between the AP and one user consists of TXSS, RXSS, SSW feedback, and SSW ACK frames, as shown in Fig. 2.9 in Chap. 2 in the monograph. Let N_t and N_r denote the number of sectors of the AP and each user, respectively. The TXSS and RXSS frames contain N_t and N_r SSW frames, respectively. With simple addition, the total consumed time in the SLS phase is given by

$$T_{SLS} = N_t T_{SSW} + N_r T_{SSW} + T_{SSW-feedback} + T_{SSW-ACK}, \qquad (4.39)$$

where T_{SSW}, $T_{SSW-feedback}$, and $T_{SSW-ACK}$ represent the time consumption for transmitting an SSW frame, an SSW feedback frame, and an SSW ACK frame, respectively.

- *ATI phase for one user*: As shown in Fig. 4.17, the ATI phase for one user consists of polling, SPR, and Grant frames. In 802.11ad, only one user can be served at each time slot in the ATI phase. Thus, we have one polling frame, one SPR frame, and one Grant frame. The total consumed time in the ATI phase is given by

$$T_{ATI} = T_{polling} + T_{SPR} + T_{Grant}, \qquad (4.40)$$

where $T_{polling}$, T_{SPR}, and T_{Grant} denote the consumed time for transmitting a polling frame, an SPR frame, and a Grant frame, respectively.

- *BRP phase for one user*: As shown in Fig. 2.10 in Chap. 2, the BRP phase consists of four subphases, i.e., BRP setup, multiple sector ID (MID), beam combining (BC), and BRP transaction subphases. The MID and BC subphases are iterative BRP training. There are three kinds of BRP frames, i.e., BRP, BRP feedback, and BRP with TRN-T/R frames. The BRP feedback frame is the BRP frame appended with channel measurement information. The BRP with a TRN-R/T frame is the BRP frame appended with the TRN-R/T field to perform channel measurement. In 802.11ad, as shown in Fig. 2.10, the components of each subphase are detailed as follows:

  - The BRP setup subphase includes five BRP frames.
  - The MID subphase includes two BRP with TRN-R/T frames and two BRP feedback frames.
  - The BC subphase includes $2N_{beam}$ BRP with TRN-R/T frames and two BRP feedback frames.
  - The BRP transaction subphase includes one BRP frame and five BRP with TRN-R/T field frames.

With simple calculation of four subphases, the total time in BRP is given by

$$T_{BRP} = T_{BRP-setup} + N_{BRP}(T_{MID} + T_{BC}) + T_{BRP-tran}, \qquad (4.41)$$

where N_{BRP} represents the number of iterations in the BRP phase.

Overall, the total beamforming training time in 802.11ad for multiple users can be calculated by the summation of SLS, ATI, and BRP phases, which is given by

$$T^s_{overhead} = N (T_{SLS} + T_{ATI} + T_{BRP}). \qquad (4.42)$$

where N is the number of the supported users.

4.6.4.2   Beamforming Training Time Analysis of Proposed Protocol

As detailed in the proposed multiuser beamforming training protocol, the protocol consists of multiuser SLS, multiuser ATI, and multiuser BRP phases. The detailed analysis of the corresponding phases is given as follows:

- *Multiuser SLS phase*: As shown in Fig. 4.16, we use multiple RXSS, SSW feedback, and SSW ACK frames to support N STAs in the coverage of the AP. The total consumed time in the multiuser SLS phase is given by

$$T_{SLS}^m = N_t T_{SSW} + N(N_r T_{SSW} + T_{SSW-feed} + T_{SSW-ACK}). \tag{4.43}$$

- *Multiuser ATI phase*: As shown in Fig. 4.17, we consider N STAs in the coverage of AP. As such, N STAs are polled by the AP, and only U users are granted with the service periods (SPs). The selected U users transmit data simultaneously in the SPs. The total consumed time in the multiuser ATI phase is

$$T_{ATI}^m = N \left(T_{polling} + T_{SPR} \right) + T_{Grant} U. \tag{4.44}$$

- *Multiuser BRP phase*: For the multiuser BRP phase, BRP is performed for each user at each slot, and hence the total consumed time is given by

$$T_{BRP}^m = T_{BRP} U. \tag{4.45}$$

Above equation illustrates that time consumption of the multiuser BRP phase increases linearly with the number of the supported users. As BRP occupies most of the beamforming training time, it implies that consumed time of the multiuser beamforming training protocol increases nearly linearly with the number of the supported users.

Overall, the total overhead of beamforming training in our proposed multiuser beamforming training protocol is given by

$$T_{overhead}^m = T_{SLS}^m + T_{ATI}^m + T_{BRP}^m. \tag{4.46}$$

Theoretical analysis results are presented via simulations to evaluate the performance of the proposed protocol, as detailed in Sect. 4.7.

4.7   Protocol Performance Evaluation

In the following, we first present the simulation setup in Sect. 4.7.1 and then illustrate the performance of the proposed multiuser beamforming training protocol in terms of the number of users and AP's transmit power in Sect. 4.7.2.

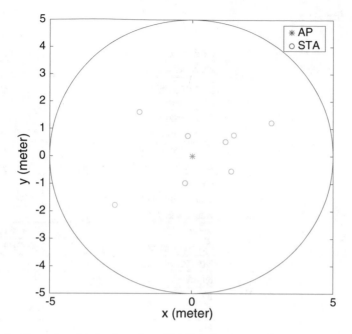

Fig. 4.19 An illustrative simulated topology in which the AP provides communication services for multiple STAs

4.7.1   Simulation Setup

We consider an indoor scenario, in which the AP has a coverage area within a radius of $R = 5$ m. The simulated topology is illustrated in Fig. 4.19, in which multiple STAs are uniformly distributed in AP's coverage. Among multiple STAs, a few users are randomly selected to perform hybrid beamforming for simultaneous data transmission. For the path loss model, we adopt a line-of-sight (LOS) link channel model. Specifically, the path loss (PL) between the AP and an STA is represented by

$$PL(dB) = A + 20 \log_{10} f + 10\gamma \log_{10} d, \qquad (4.47)$$

where $A = 32.5$, d is the distance between AP and the STA, f is the central carrier frequency (in GHz), and $\gamma = 2$ [44]. The duration of a BI is set to 1000 ms in the simulation, as suggested by 802.11ad. The AP is equipped with 16 transmitter sectors, and each STA is equipped with 8 receiver sectors. The other important simulation parameters are listed in Table 4.3.

Table 4.3 Simulation
parameters

Parameter	Value
SSW frame size	26 bytes
SSW feedback frame size	28 bytes
SSW ACK frame size	28 bytes
Polling frame size	22 bytes
SPR frame size	27 bytes
Grant frame size	27 bytes
BRP frame size	986 bytes
BRP feedback frame size	1008 bytes
BRP with TRN-R/T frame size	2830 bytes
Control link rate	27.5 Mbps
Beacon interval duration	1000 ms
Background noise spectrum density	-134 dBm/MHz
Bandwidth	2.16 GHz
Central carrier frequency	60 GHz
Number of antennas at AP	16
Number of antennas at user	8
N_t	16
N_r	8
N_{BRP}	4
N_{beam}	4
N	8
U	4
R	5 m

4.7.2   Simulation Results

As shown in Fig. 4.20, we show the overhead performance of the 802.11ad beamforming training protocol with respect to different numbers of users. Theoretical
overhead analysis for the 802.11ad beamforming training protocol that supports
single user transmission is given in (4.42). The theoretical overhead analysis of
the proposed hybrid beamforming protocol is given in (4.46). We can see that the
protocol overhead increases linearly with the number of users. The reason is that
multiuser BRP stage is implemented at different time slots for different users. For
a hybrid beamforming case with eight users, the beamforming training overhead
is about 400 ms, which is expected to occupy about 40% of the entire duration of
an BI in 802.11ad. Simulation results indicate that the current hybrid beamforming
protocol in 802.11ad is inefficient to support a large number of users.

Taking the overhead of beamforming training into consideration, we aim to
compare the effective data rate performance in the following. Firstly, we define the
effective rate as

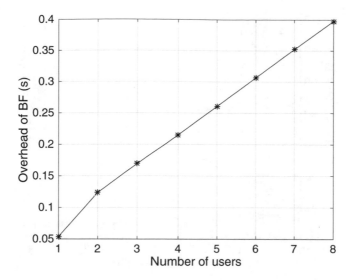

Fig. 4.20 Time consumption of beamforming training in 802.11ad standard in terms of different numbers of users

$$R_e = \frac{(T - T_{overhead})R_{sum}}{T}$$

where R_e is the effective rate, $T_{overhead}$ is the time consumption of beamforming training overhead, and R_{sum} is the raw transmission rate calculated by (4.35). Secondly, as shown in Fig. 4.21, the effective date rate performance of hybrid beamforming with respect to different transmitting powers is compared. Simulation results demonstrate that the proposed hybrid beamforming protocol can achieve a significant data rate gain compared with the single user transmission in 802.11ad standard. This is because that hybrid beamforming can exploit multiuser multiplexing gain to enhance throughput and support multiuser transmission. However, as the number of users that performs hybrid beamforming increases, the overhead of beamforming training increases significantly. In the figure, the AP's transmission power is set to 0.001 mW, and a five-user hybrid beamforming scheme outperforms the others. However, the result does not hold in other cases. As such, it is very important to identify the optimal number of users to perform hybrid beamforming with different AP's transmitting powers, thereby enhancing system performance.

As shown in Fig. 4.22, we compare the effective rates of the proposed hybrid beamforming protocols with different numbers of users when transmitting powers are set to 0.001, 0.1, and 10 mW, respectively. Simulation results show the optimal number of users varies in terms of transmit powers. Specifically, when the transmitting power is low, such as 0.001 mW, we can see that the optimal number of users to perform hybrid beamforming is 5. Then, when the transmitting power is 0.1 mW,

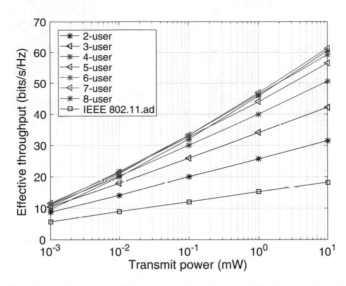

Fig. 4.21 Effective sum rate comparison of the proposed hybrid beamforming protocol with different numbers of users

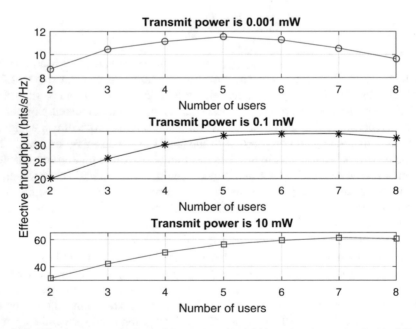

Fig. 4.22 Effective rate comparison of the proposed hybrid beamforming protocol with different numbers of users and transmitting powers

the optimal number of users changes to 6. Next, with a higher transmitting power, such as 10 mW, the optimal number of users is 7. The results demonstrate that the optimal number of users increases with the AP's transmitting power strength.

4.8 Summary

In this chapter, we focus on the beamforming training protocol design and performance analysis in both single-user transmission and multiuser transmission scenarios. For the single-user transmission scenario, we have introduced the BFT-MAC performance analysis and enhancement in the current 802.11ad. An accurate analytical model has been introduced to analyze the performance of BFT-MAC, and the asymptotic analysis has demonstrated that the maximum normalized throughput is barely $1/e$. We have introduced an enhancement scheme to enhance the performance in dense user scenarios. For the multiuser transmission scenario, we have introduced a multiuser beamforming training protocol. Furthermore, we have analyzed the overhead of our protocol and found that overhead increases nearly linearly with the number of the simultaneously supported users.

References

1. W. Wu, N. Cheng, N. Zhang, P. Yang, K. Aldubaikhy, X. Shen, Performance analysis and enhancement of beamforming training in 802.11ad. IEEE Trans. Veh. Technol. **69**(5), 5293–5306 (2020)
2. W. Wu, Q. Shen, M. Wang, X. Shen, Performance analysis of IEEE 802.11.ad downlink hybrid beamforming, in *Proceedings of IEEE International Conference on Communications (ICC)* (2017), pp. 1–6
3. W. Wu, Q. Shen, K. Aldubaikhy, N. Cheng, N. Zhang, X. Shen, Enhance the edge with beamforming: performance analysis of beamforming-enabled WLAN, in *Proceedings of IEEE 16th International Symposium on Modeling and Optimization in Mobile, Ad Hoc, and Wireless Networks (WiOpt)* (2018), pp. 1–6
4. S. Sur, X. Zhang, P. Ramanathan, R. Chandra, BeamSpy: enabling robust 60 GHz links under blockage, in *Proceedings of Proceedings of the 13th USENIX Symposium on Networked Systems Design and Implementation (NSDI'16)* (2016), pp. 193–206
5. W. Wu, N. Zhang, N. Cheng, Y. Tang, K. Aldubaikhy, X. Shen, Beef up mmwave dense cellular networks with D2D-assisted cooperative edge caching. IEEE Trans. Veh. Technol. **68**(4), 3890–3904 (2019)
6. J. Wang, Z. Lan, C. Pyo, T. Baykas, C. Sum, M.A. Rahman, J. Gao, R. Funada, F. Kojima, H. Harada, S. Kato, Beam codebook based beamforming protocol for multi-Gbps millimeter-wave WPAN systems. IEEE J. Sel. Areas Commun. **27**(8), 1390–1399 (2009)
7. Z. Marzi, D. Ramasamy, U. Madhow, Compressive channel estimation and tracking for large arrays in mm-wave picocells. IEEE J. Sel. Topics Signal Process. **10**(3), 514–527 (2016)
8. S. Sur, I. Pefkianakis, X. Zhang, K.H. Kim, WiFi-assisted 60 GHz wireless networks, in *Proceedings of ACM International Conference on Mobile Computing and Networking (MOBICOM)* (2017), pp. 28–41
9. X. Shen, J. Gao, W. Wu, K. Lyu, M. Li, W. Zhuang, X. Li, J. Rao, AI-assisted network-slicing based next-generation wireless networks. IEEE Open J. Veh. Technol. **1**(1), 45–66 (2020)

10. W. Wu, N. Chen, C. Zhou, M. Li, X. Shen, W. Zhuang, X. Li, Dynamic RAN slicing for service-oriented vehicular networks via constrained learning. IEEE J. Sel. Areas Commun. **39**(7), 2076–2089 (2021)
11. G.H. Sim, S. Klos, A. Asadi, A. Klein, M. Hollick, An online context-aware machine learning algorithm for 5G mmWave vehicular communications. IEEE/ACM Trans. Netw. **26**(6), 2487–2500 (2018)
12. W. Wu, N. Cheng, N. Zhang, P. Yang, W. Zhuang, X. Shen, Fast mmwave beam alignment via correlated bandit learning. IEEE Trans. Wireless Commun. **18**(12), 5894–5908 (2019)
13. P. Zhou, X. Fang, Y. Fang, Y. Long, R. He, X. Han, Enhanced random access and beam training for millimeter wave wireless local networks with high user density. IEEE Trans. Wireless Commun. **16**(12), 7760–7773 (2017)
14. A. Alkhateeb, G. Leus, R.W. Heath, Limited feedback hybrid precoding for multi-user millimeter wave systems. IEEE Trans. Wireless Commun. **14**(11), 6481–6494 (2015)
15. F. Sohrabi, W. Yu, Hybrid digital and analog beamforming design for large-scale antenna arrays. IEEE J. Sel. Topics Signal Process. **10**(3), 501–513 (2016)
16. A. Alkhateeb, G. Leus, R.W. Heath, Limited feedback hybrid precoding for multi-user millimeter wave systems. IEEE Trans. Wireless Commun. **14**(11), 6481–6494 (2015)
17. H. Hassanieh, O. Abari, M. Rodriguez, M. Abdelghany, D. Katabi, P. Indyk, Fast millimeter wave beam alignment, in *Proceedings of Proceedings of the 2018 Conference of the ACM Special Interest Group on Data Communication* (2018), pp. 432–445
18. G. Bianchi, Performance analysis of the IEEE 802.11 distributed coordination function. IEEE J. Sel. Areas Commun. **18**(3), 535–547 (2000)
19. T.H. Luan, X. Ling, X. Shen, MAC in motion: impact of mobility on the MAC of drive-thru Internet. IEEE Trans. Mobile Comput. **11**(2), 305–319 (2012)
20. W. Xu, W. Shi, F. Lyu, H. Zhou, N. Cheng, X. Shen, Throughput analysis of vehicular Internet access via roadside WiFi hotspot. IEEE Trans. Veh. Technol. **68**(4), 3980–3991 (2019)
21. Q. Ye, W. Zhuang, L. Li, P. Vigneron, Traffic-load-adaptive medium access control for fully connected mobile ad hoc networks. IEEE Trans. Veh. Technol. **65**(11), 9358–9371 (2016)
22. Q. Ye, W. Zhuang, L. Li, P. Vigneron, Traffic-load-adaptive medium access control for fully connected mobile ad hoc networks. IEEE Trans. Veh. Technol. **65**(11), 9358–9371 (2016)
23. X. Yuan, C. Li, Q. Ye, K. Zhang, N. Cheng, N. Zhang, X. Shen, Performance analysis of IEEE 802.15.6-based coexisting mobile WBANs with prioritized traffic and dynamic interference. IEEE Trans. Wireless Commun. **17**(8), 5637–5652 (2018)
24. K. Chandra, V. Prasad, I. Niemegeers, Performance analysis of IEEE 802.11ad MAC protocol. IEEE Commun. Lett. **21**(7), 1513–1516 (2017)
25. Q. Chen, J. Tang, D.T.C. Wong, X. Peng, Y. Zhang, Directional cooperative MAC protocol design and performance analysis for IEEE 802.11ad WLANs. IEEE Trans. Veh. Techol. **62**(6), 2667–2677 (2013)
26. S. Shao, H. Zhang, D. Koutsonikolas, A. Khreishah, Two-dimensional reduction of beam training overhead in crowded 802.11ad based networks, in *Proceedings of IEEE INFOCOM Workshop* (2018), pp. 680–685
27. S.G. Kim, K. Jo, S. Park, H. Cho, J. Kim, S. Bang, On random access in A-BFT. IEEE 802.11 Documents, doc.:IEEE 802.11-16/0948-00-00ay (2016)
28. Y. Xin, R. Sun, O. Aboul-Magd, Channel access in A-BFT over multiple channels. IEEE 802.11 Documents, doc.:IEEE 802.11-16/0101r0 (2016)
29. K. Jo, S. Park, H. Cho, J. Kim, S. Bang, S.G. Kim, Short SSW frame for A-BFT. IEEE 802.11 Documents, doc.:IEEE 802.11-17/0117-00-00ay (2017)
30. Y. Ghasempour, C.R. da Silva, C. Cordeiro, E.W. Knightly, IEEE 802.11 ay: next-generation 60 GHz communication for 100 Gb/s Wi-Fi. IEEE Commun. Mag. **55**(12), 186–192 (2017)
31. K. Aldubaikhy, W. Wu, N. Zhang, N. Cheng, X. Shen, mmwave IEEE 802.11ay for 5G fixed wireless access. IEEE Wireless Commun. **27**(2), 88–95 (2020)
32. The IEEE Standards, IEEE standards 802.11 ad-2012: Enhancement for very high throughput in the 60 GHz band (2012)

33. P. Zhou, K. Cheng, X. Han, X. Fang, Y. Fang, R. He, Y. Long, Y. Liu, IEEE 802.11ay-based mmWave WLANs: design challenges and solutions. IEEE Commun. Surveys Tuts. **20**(3), 1654–1681 (2018)
34. N. Zhang, N. Cheng, A.T. Gamage, K. Zhang, J.W. Mark, X. Shen, Cloud assisted hetnets toward 5G wireless networks. IEEE Commun. Mag. **53**(6), 59–65 (2015)
35. R.W. Heath, N. Gonzalez-Prelcic, S. Rangan, W. Roh, A.M. Sayeed, An overview of signal processing techniques for millimeter wave MIMO systems. IEEE J. Sel. Topics Signal Process. **10**(3), 436–453 (2016)
36. R. Méndez-Rial, C. Rusu, N. González-Prelcic, A. Alkhateeb, R.W. Heath, Hybrid MIMO architectures for millimeter wave communications: phase shifters or switches? IEEE Access **4**, 247–267 (2016)
37. A. Alkhateeb, R.W. Heath, Frequency selective hybrid precoding for limited feedback millimeter wave systems. IEEE Trans. Commun. **64**(5), 1801–1818 (2016)
38. C. Rusu, R. Méndez-Rial, N. González-Prelcicy, R.W. Heath, Low complexity hybrid sparse precoding and combining in millimeter wave MIMO systems, in *Proceedings of IEEE International Conference on Communications (ICC)* (2015), pp. 1340–1345
39. A. Alkhateeb, O. El Ayach, G. Leus, R.W. Heath, Hybrid precoding for millimeter wave cellular systems with partial channel knowledge, in *Proceedings of IEEE Information Theory and Applications Workshop (ITA)* (2013), pp. 1–5
40. M. Gao, B. Ai, Y. Niu, W. Wu, P. Yang, F. Lyu, X. Shen, Efficient hybrid beamforming with anti-blockage design for high-speed railway communications. IEEE Trans. Veh. Technol. **69**(9), 9643–9655 (2020)
41. K. Aldubaikhy, W. Wu, Q. Ye, X. Shen, Low-complexity user selection algorithms for multiuser transmissions in mmwave WLANs. IEEE Trans. Wireless Commun. **19**(4), 2397–2410 (2020)
42. K. Aldubaikhy, W. Wu, X. Shen, HBF-PDVG: hybrid beamforming and user selection for UL MU-MIMO mmWave systems, in *Proceedings of IEEE Globecom Workshop* (2018), pp. 1–6
43. J. Qiao, X. Shen, J.W. Mark, Y. He, MAC-layer concurrent beamforming protocol for indoor millimeter-wave networks. IEEE Trans.Veh.Technol. **64**(1), 327–338 (2014)
44. V.E.A. Maltsev, Channel models for 60 GHz WLAN systems. doc.: IEEE 802.11-09/0334r8, no. 3

Chapter 5
Beamforming-Aided Cooperative Edge Caching in mmWave Dense Networks

5.1 Introduction

The surging data-intensive mobile applications, e.g., virtual reality, mobile augmented reality, and high-definition immersive video streaming, are constantly challenging the capacity of wireless networks, driving the network evolution to the next generation [1–3]. To properly provision the ever-increasing demands of mobile data traffic, mmWave communication, which is a de facto candidate technology for the fifth-generation (5G) networks, is envisaged to provide a pseudo-wire wireless connection service by exploiting a large swath of spectrum resource [4]. Leveraging high-gain directional antennas, current mmWave networks offer an extremely high data rate of nearly 7 Gbps, which is expected to increase to 40 Gbps in the forthcoming future [5]. As mmWave networks can be further densified due to the hostile propagation characteristics, deploying unconstrained wired backhaul links in dense networks becomes infeasible due to high costs, which results in backhaul congestion. Alleviating backhaul pressure is imperative for mmWave dense networks in the future.

Edge caching, which harnesses the feature of repetitive content requests of mobile applications, can effectively relieve the backhaul pressure [6–8]. Specifically, popular contents that are being frequently requested can be cached on the edge nodes (e.g., small base station (SBS) [9]) during off-peak hours to serve neighboring users during peak hours. In this way, the backhaul traffic is expected to be reduced by up to 35% [10, 11]. Meanwhile, edge caching also helps reduce content deliver delay as the content is retrieved from the edge instead of remote servers. However, due to the limited capacity of each edge node, it is rather difficult to boost the performance of edge caching. To achieve a higher edge caching gain, an intuitive solution is to employ caching resources on different edge nodes in a cooperative manner, i.e., cooperative caching. Cooperative caching can be divided into two categories: (a) *cooperative edge caching* where contents are cached in the SBS cluster consisting of multiple SBSs in proximity and (b) *device-to-device (D2D) caching* where contents

P. Yang et al., *Millimeter-Wave Networks*, Wireless Networks,
https://doi.org/10.1007/978-3-030-88630-1_5

are cached in nearby users. In the former, each SBS in the cluster caches a diverse set of contents to serve users with an increased caching diversity, while in the latter case, each user and its neighboring user cache contents and exchange cached contents via high-rate D2D communications.

Implementing cooperative caching in mmWave dense networks can help substantially reduce the backhaul burden and suppress the content retrieval delay. Meanwhile, it also brings extra opportunity in tackling the interference issue. Specifically, in a wireless system operating in lower-frequency bands, the performance of cooperative caching is limited by multiuser interference, as omnidirectional antennas are used for data transmission. In contrast, mmWave systems employ directional antennas, which naturally avoid interference from neighboring users. Yet, cooperative caching in mmWave networks also confronts new challenges. Firstly, there exists no well-established analytical model for mmWave networks, clarifying the relationship between network density and content caching in case directional antennas are used. As a consequence, it is oftentimes difficult to achieve a closed-form expression that gives intuitive insights for system design.

In this chapter, to address the aforementioned challenges, we devise a new cooperative caching policy for mmWave cellular networks. This policy exploits the cooperative caching resources from users and SBSs in their proximity. The content caching decisions are thus made by accounting for the content retrieval delay. Intuitively, the most popular contents can be cached in the users and its D2D peers, in order to minimize the content retrieval delay, while less popular contents can be cached in the SBS cluster. In light of such caching policy, we analyze the backhaul offloading gain by considering a directional mmWave antenna. Such antenna is featured with varying main lobe antenna gain and non-zero side lobe gain, which is the key challenge of interference analysis. By harnessing the stochastic network topology information, the average content retrieval delay of the introduced policy is theoretically analyzed. The results of which reveal how the network density and directional antennas affect the overall caching performance.

The remainder of this chapter is organized as follows. Section 5.2 reviews related works, followed by the system model in Sect. 5.3. We present the DCEC design and analyze the corresponding backhaul offloading gain in Sect. 5.4. Section 5.5 gives theoretical performance analysis on the content retrieval delay. Extensive simulations are provided in Sect. 5.6. Finally, concluding remarks are given in Sect. 5.7.

5.2   Related Works on Edge Caching

Equipped with both computing and storage capabilities, mobile edge computing (MEC) [12] is capable of providing high quality of experience (QoE) to mobile users. Recently, extensive efforts have been devoted to the computing functionality of MEC [13–15]. In order to minimize the service latency, Rodrigues et al. devised a hybrid method consisting of virtual machine migration and transmission power control [13]. Based on this method, they proposed to reconfigure edge servers to account for system scalability [14]. Zhou et al. considered to jointly optimize the allocation of computing and caching resources in mobile edge networks, in order to maximize the system utility [15].

The caching resources on the edge nodes provide another approach to enhance user's QoE [16, 17]. Generally, existing literature on edge caching can be classified into two categories: D2D caching and cooperative edge caching. Both of them have been extensively studied in case of microwave communications. For instance, it is found that by leveraging both the user's caching resources and the high-rate D2D communications [18–20], D2D caching is able to effectively offload cellular traffic, hence improving cellular data rate and reducing power consumption. Theoretic scaling law also indicates that, if the D2D transmission range can be made adaptive to the network density, the network throughput would increase with the number of network nodes [21]. Wang et al. focused on a mobile scenario, where users frequently interact with others in their proximity via D2D communications [22]. In [23], three scheduling schemes in D2D communication-based edge caching are proposed, each of which improves the throughput of D2D links at low computational complexity.

Cooperative edge caching aims at exploiting the caching resources within the SBS cluster to enhance the capability of local caching. Chen et al. devised a cooperative caching policy by caching different portions of less popular contents at different SBSs to increase content diversity. They then studied the tradeoff relationship between transmission diversity and content diversity [24]. In order to maximize the caching performance, both content placement and caching capacity have been optimized. Zhang et al. investigated the delay-optimal caching problem by means of content placement, and a greedy algorithm is proposed to achieve the goal of cooperative edge caching [25]. Another work addresses the caching size optimization problem considering the budget of cache deployment in heterogeneous networks [26]. Recently, the authors of [27] proposed to apply in-memory storage and processing to enhance the energy efficiency of edge caching. Zhao et al. developed a cooperative caching policy according to the distribution of content popularity and user preference [28], leading to improved content hit ratio and reduced transmission delay. Furthermore, observing the fact that content popularity is highly dependent on user locations, Yang et al. presented a location-aware caching strategy by adaptive learning of content popularity [29].

In summary, the aforementioned works focus on cooperative caching policies in case of microwave communications, whereas the unique capability and feature of mmWave communications in wireless networks are seldom considered. Particularly, the mutual interference of mobile users with omni-directional antennas poses significant challenge for at microwave bands, especially in ultradense networks. Consequently, complex interference cancellation and interference management techniques are required for both D2D caching and cooperative edge caching [25, 30]. Fortunately, such interference issue can be readily addressed using directional antennas in mmWave communications. In light of D2D edge caching using mmWave communications, Semiari et al. focused on the handover failures in mobile mmWave networks and proposed a proactive caching policy [31]. Ji et al. proposed to use D2D caching in mmWave networks to improve network performance [32]. However, analytical results are not provided to feature the performance of D2D caching. Giatsoglou et al. proposed a D2D caching policy and leveraged the stochastic geometry theory for performance analysis [33]. However, the cache resource on SBSs is not included in the system. The impact of directional mmWave antennas is not characterized either. The work of both [32] and [33] only applies D2D mmWave links for backhaul traffic offloading, without considering the impact of network density. To address those issues, this chapter focuses on cooperative edge caching and takes into account of the high data rate of mmWave communications to further improve the caching performance.

5.3　System Model

In this section, we first present the network model and the content popularity model. Then, we characterize the directional antenna and the mmWave channel, followed by the data transmission model. Important notations are summarized in Table 5.1.

Table 5.1 Important notations

Notation	Description
$\mathcal{F}$	Requested file library
Φ_{BS}	PPP of SBS
λ_{BS}	Density of SBS
Φ_{UE}	PPP of users
λ_{UE}	Density of users
W	System bandwidth
ϕ	Fraction of bandwidth allocation
α	Path loss exponent
K	SBS cluster size
r	Physical distance
R	Transmission rate
σ^2	Background noise power
D	Average content retrieval delay
S	Signal power
I	Interference power
ξ	Content popularity skewness
G	Directional antenna gain
F	Backhaul offloading gain
h	Content hit ratio
$\mathcal{B}_o$	Associated SBS

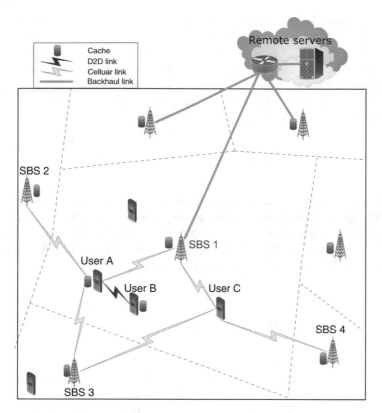

Fig. 5.1 Cache-enabled edge network topology

5.3.1 Network Model

Figure 5.1 gives an illustration of the topology of a cache-enabled edge network. Without loss of generality, we assume the geographical distribution of SBSs and users follows homogeneous Poisson point process (PPP) Φ_{BS} and Φ_{UE}, respectively, the density of which are λ_{BS} and λ_{UE} [33, 34]. Each SBS operates on a shared spectrum band and communicates to remote servers via a backhaul link with limited capacity. Time division multiple access (TDMA) mode is adopted by each SBS to serve mobile users in its coverage. Both SBSs and users use steerable directional antennas for packet transmission. We assume beamforming training is perfectly performed between users and associated SBSs before data transmission [35, 36].

We consider a user-centric architecture [25], each user can be served by at most K SBSs, which forms a SBS cluster represented by $\{SBS_1, SBS_2, \ldots, SBS_K\}$. For instance, Fig. 5.1 shows user A is served by a SBS cluster of three SBSs, $\{SBS_1, SBS_2, SBS_3\}$. Users are divided into two types: unpaired users and paired users. For unpaired users, they follow a homogeneous PPP Φ_u with parameter λ_u.

Such users can only be served by SBSs (e.g., user C in Fig. 5.1). By contrast, paired users can be served by not only SBS cluster but also its D2D peer via high-rate D2D communications. Let Φ_p be the homogeneous PPP followed by paired users, with a density of λ_p. User A and B form a D2D pair and are able to communicate with each other via a D2D link. We assume the D2D peer of a paired user locates uniformly within a disk of radius r_d^{max}. Thus, the distance, r_d, between the user and its corresponding D2D peer is given by the following distribution [33]:

$$f(r_d) = \frac{2r_d}{\left(r_d^{max}\right)^2}, 0 < r_d < r_d^{max}. \tag{5.1}$$

Note that both D2D communications and cellular communications are simultaneously supported in the system; the overlay scheme is employed. That is, D2D communications and cellular communications are using disjoint frequency bands to avoid interference. Denote by W the available bandwidth for the mmWave system, while ϕW is the bandwidth allocated to D2D communications.

5.3.2 Content Popularity Model

Denote by $\mathcal{F} = \{f_1, f_2, \ldots, f_i, \ldots f_{|\mathcal{F}|}\}$ and $Q = \{q_1, q_2, \ldots, q_i, \ldots, q_{|\mathcal{F}|}\}$ the sets of contents that can be requested and their corresponding popularity, respectively, where $|\mathcal{F}|$ is the number of contents. We assume the popularity distribution follows a Zipf distribution [25], and the popularity of the i-th popular content is given by

$$q_i = \frac{i^{-\xi}}{\sum_{j=1}^{|\mathcal{F}|} j^{-\xi}}, 1 \le i \le |\mathcal{F}| \tag{5.2}$$

where $\xi \ge 0$ is the skewness of the content popularity distribution. The skewness constant varies with content types, and a larger value of skewness indicates that the popularity distribution is more concentrated.

5.3.3 Directional Antenna Model

The ideal "flat-top" model has been widely adopted in the literature. This model is featured with a constant gain in the main lobe and zero-gain elsewhere, which helps largely simplify the interference analysis [37]. In practice, however, the main-lobe gain of directional antenna varies and the side-lobe gain is non-zero. In this chapter, the practical antenna model is adopted, the gain of which is characterized according to the relative angle, θ, to its boresight, i.e.,

$$G(\theta) = \begin{cases} G_m 10^{-c\left(\frac{2\theta}{\omega_m}\right)^2} & |\theta| \leq \frac{\theta_m}{2} \\ G_s & \frac{\theta_m}{2} < |\theta| \leq \pi. \end{cases} \tag{5.3}$$

G_m and G_s are the maximum antenna gain of the main lobe and the average antenna gain of the side lobe, respectively. ω_m and θ_m represent the beam width of the half-power and main lobe, respectively. c is a constant, the empirical value of which is 0.3 [2].

5.3.4 mmWave Channel Model

For the mmWave channel model, we consider the large-scale fading of mmWave links, which is modeled as

$$PL(dB) = 20 \log_{10}\left(\frac{4\pi d_0}{\lambda}\right) + 10\alpha \log_{10}\left(\frac{r}{d_0}\right), r \geq d_0 \tag{5.4}$$

where r is the propagation distance and α is the path loss exponent, λ represents the wavelength, and d_0 is the free space reference distance [38, 39]. This model can well characterize the mmWave channel when the propagation distance is further than the reference distance. For notation simplicity, the average path loss is rewritten as

$$\beta = Cr^{-\alpha} \tag{5.5}$$

where $C = \frac{\lambda^2}{d_0^3(4\pi)^2}$ is a constant.

In this chapter, the fast Raleigh fading is utilized for characterizing the small-scale fading, i.e., $h \sim \exp(1)$, which means that the channel gain is an exponential random variable with a unit mean.

5.3.5 Transmission Model

The transmission throughput of a mmWave link can be obtained by

$$R = \frac{W}{N_{cell}} \log_2\left(1 + \frac{S}{I + \sigma^2}\right) \tag{5.6}$$

where N_{cell} is the cell load and σ^2 is the power of background noise given by the noise power spectral density N_o and bandwidth W, i.e., $\sigma^2 = WN_o$. S and I are the power of signal and interference, respectively.

Each user receives interference from potentially all the SBSs except its associated SBS, $\mathcal{B}_o$. With the model of both the directional antenna and the channel, the received interference power is given by

$$
\begin{aligned}
I &= \sum_{i \in \Phi_{BS} \backslash \mathcal{B}_o} I_i \\
&= \sum_{i \in \Phi_{BS} \backslash \mathcal{B}_o} P_B G(\theta_{t,i}) G(\theta_{r,i}) h_i C r_i^{-\alpha}
\end{aligned}
\tag{5.7}
$$

where $G(\theta_{t,i})$ and $G(\theta_{r,i})$ are the gain of transmitting and receiving directional antenna, respectively. $\theta_{t,i}$ and $\theta_{r,i}$ are the angle of departure (AOD) and the angle of arrival (AOA) of the interference signal between the user and the ith interfered SBS, respectively. r_i is the physical distance between the interfered user and the ith SBS. For analytical simplicity, the AOAs and AODs of interference links are assumed to be uniformly distributed in $(0, 2\pi]$ [40], which gives an approach to estimate the average directional antenna gain of the interference signal, i.e.,

$$
\begin{aligned}
\bar{G} &= \int_0^{2\pi} G(\theta) f(\theta) d\theta \\
&= \int_0^{\frac{\theta_m}{2}} G_m 10^{-c \left(\frac{2\theta}{\omega_m} \right)^2} \frac{1}{\pi} d\theta + \int_{\frac{\theta}{2}}^{\pi} G_s \frac{1}{\pi} d\theta \\
&= \frac{\omega_m G_m}{\sqrt{2c\pi \ln 10}} \text{erfc} \left(\frac{\theta_m \sqrt{c \ln 10}}{\omega_m} \right) - \frac{G_s \theta_m}{2\pi} + G_s
\end{aligned}
\tag{5.8}
$$

where $\text{erfc}(x)$ is the Gauss error function equals $\int_0^x e^{-t^2} dt$. Then, the average interference power in (5.7) can be rewritten as

$$
\mathbb{E}[I] = \sum_{i \in \Phi_{BS} \backslash \mathcal{B}_o} P_B \bar{G}^2 C \mathbb{E}[h_i] \mathbb{E}[r_i^{-\alpha}]
\tag{5.9}
$$

which is used in the following analysis of this chapter.

5.4   D2D-Assisted Cooperative Edge Caching (DCEC) Policy

In this section, the DCEC policy is firstly proposed with the aim of exploiting caching diversity, and then the backhaul offloading performance is theoretically analyzed.

5.4.1 Scheme Design

The key idea of the DCEC policy is to utilize the pooling cache resource from mobile users, their D2D peers, and the SBS cluster, to store popular contents, which helps reduce the backhaul traffic. Note that it is the user who requests its D2D peer and the SBS cluster to obtain the caching information of interested contents. For a requested content, if it is cached at the user end, it can be retrieved by the user at negligible latency. If the content is cached in its D2D peer, it can be retrieved by the users via D2D communications. Last, if the content is cached in a certain SBS, the user can connect to the SBS and request the content via cellular links, at the cost of a higher delay than that via D2D communications. This is because D2D communications operate at mmWave band and, meanwhile, the D2D peers locate at a shorter distance than the SBS, which help provide much higher data rates. In contrast, if the content is not cached locally, the user connects to the nearest SBS B_o and then the remote servers to retrieve the contents, which incurs a large and unpredictable delay. In this way, for each mobile user, the place for content retrieval is prioritized based on the delay in an ascending order, i.e., {user end $\leq$ its D2D peer $\leq$ SBS cluster $\leq$ remoter servers}. Such priority sequence is referred during the content placement of the proposed DCEC policy to minimize content retrieval delay.

Firstly, in order to provide content retrieve services at minimum delay, the most popular contents are cached on both the user end and its D2D peer. Without loss of generality, we assume each user is equipped with the same storage capacity, C_u. Recall that there are paired users and unpaired users. For the former, a set of the $2C_u$ most popular contents, i.e., $\{f_1, f_2, \ldots, f_{2C_u}\}$, are cached in the D2D pair of users. For the sake of fairness, these $2C_u$ contents are evenly distributed in the user and its D2D peer such that the contents at both paired users have the same request probability. In this way, the content hit ratio of one of the paired user is

$$h_p = \frac{1}{2} \sum_{i=1}^{2C_u} q_i. \tag{5.10}$$

For unpaired users, each of them can only cache a set of the most popular C_u contents, i.e., $\{f_1, f_2, \ldots, f_{C_u}\}$. The content hit ratio at an unpaired user is

$$h_u = \sum_{i=1}^{C_u} q_i. \tag{5.11}$$

Secondly, considering the long retrieval delay via cellular communications, those less popular contents are cached in the SBS cluster. Similarly, we assume each SBS has equal cache capacity, C_s. Each user is served by a cluster consisting of K SBS. Hence, the SBS cluster caches the set of KC_s less popular contents, i.e., $\{f_{2C_u+1}, f_{2C_u+2}, \ldots, f_{2C_u+KC_s}\}$. For the sake of fairness and load balancing, these

KC_s contents are evenly distributed in each SBS of the cluster. Formally, the content hit ratio at each SBS is given by

$$h_s = \frac{1}{K} \sum_{i=2C_u+1}^{2C_u+KC_s} q_i. \tag{5.12}$$

Note that it is possible for some fair popular contents being mis-cached for the unpaired users. Yet, in the case of dense networks, the portion of unpaired users is negligible, as users can connect to proximal users with a high probability. In our subsequent simulations, the ratio of paired and unpaired users is set to be 80% and 20%, respectively.

5.4.2   Backhaul Offloading Analysis

Thanks to the proposed DCEC policy, instead of retrieving contents via capacity-constrained backhaul links, users can obtain their requested contents within edge networks, which significantly reduce the backhaul burden. Before theoretically characterizing the performance gain of the DCEC policy, we first define the *backhaul offloading gain* as the amount of data traffic that is not served by backhaul links to the total data traffic ratio.

For any two paired users, the total storage of both users and the SBS cluster can be utilized to store the most popular $2C_u + KC_s$ contents. As a result, the backhaul offloading gain is $2h_p + Kh_s$. For those unpaired users, without the mutual cooperation of D2D peers, the resulting backhaul offloading gain is $h_u + Kh_s$. Collectively, the DCEC policy enjoys an average backhaul offloading gain of

$$F = h_u(1 - \delta) + 2h_p\delta + Kh_s \tag{5.13}$$

where $\delta = \frac{\lambda_p}{\lambda_p+\lambda_u}$ is the portion of paired users of all the users. Clearly, the value of gain increases with the cluster size K, as a larger cluster size brings higher potential of caching diversity gain.

We can also obtain the corresponding content miss probability, which is the chance that the requested content is not locally cached, as

$$P_m = 1 - F. \tag{5.14}$$

According to our content priority, those contents are less popular and can be obtained from remote servers by connecting to the nearest cellular link and then the constraint backhaul link.

5.5 Content Retrieval Delay Analysis

In this section, we analyze the average *content retrieval delay* of the proposed DCEC policy. Since users may retrieve the interested content through different links, the average content retrieval delay is thus analyzed in accordance with those communication links. In case the requested content is not cached, this request is delivered to remote servers via a cellular link and a backhaul link. Note that the backhaul link is capacity-constrained, with an average transmission rate of $\mathbb{E}[R_B]$. Denote by $\mathbb{E}[R_N]$ the average transmission rate between the user and its nearest SBS. If the requested content happens to be cached in either the SBS cluster or its D2D peer, the user can retrieve the content at an average transmission rate, $\mathbb{E}[R_C]$, from the SBS cluster or at an average transmission rate, $\mathbb{E}[R_D]$, from its D2D peer. Denote by v the average content size, we can obtain the average content retrieval delay of the proposed DCEC policy as

$$D = \frac{P_m v}{\mathbb{E}[R_B]} + \frac{P_m v}{\mathbb{E}[R_N]} + \frac{P_s v}{\mathbb{E}[R_C]} + \frac{P_d v}{\mathbb{E}[R_D]} \tag{5.15}$$

where $P_s = K h_s$ and $P_d = \delta h_p$ represent the probabilities that the requested content is cached in the SBS cluster and the D2D peer, respectively. In what follows, the transmission rate in different cases is theoretically lower bounded respectively, based on which an upper bound of the average content retrieval delay can be provided.

5.5.1 Backhaul Transmission Rate Analysis

Recall that the geographical distribution of both users and the SBSs follows the PPP. According to the property of the PPP, the amount of traffic generated from each network area also follows PPP. For those users served by constraint backhaul links, they are modeled as a homogeneous PPP, Φ_B, with parameter $P_m \lambda_{UE}$. Each SBS is assumed to be with equal backhaul capacity, B, serving its associated users in a TDMA manner. The following lemma gives the average backhaul transmission rate.

Lemma 5.1 *For each user, the average backhaul transmission rate is given by*

$$\mathbb{E}[R_B] = \frac{B \lambda_{BS}}{P_m \lambda_{UE}} \frac{\left(1 + \frac{P_m \lambda_{UE}}{\kappa \lambda_{BS}}\right)^{\kappa+1}}{\left(1 + \frac{P_m \lambda_{UE}}{\kappa \lambda_{BS}}\right)^{\kappa+1} - 1} \tag{5.16}$$

where κ is a constant equal to 3.5.

Proof Since the backhaul capacity is evenly allocated to each user, the resulting average backhaul rate of each user is $B/\mathbb{E}[N_B]$, where N_B is a random variable depending on the SBS cell area and characterizes the backhaul load. Without loss of generality, we assume the SBS cell area, a, follows a Gamma distribution with parameter κ [41], and the probability distribution function (PDF) is given by

$$f(a) = a^{\kappa-1} e^{-\kappa \lambda_{BS} a} \frac{(\kappa \lambda_{BS})^{\kappa}}{\Gamma(\kappa)}. \tag{5.17}$$

Then, the average backhaul load can be given by

$$
\begin{aligned}
\mathbb{E}[N_B] &= \int_a^{\infty} \sum_{n=1}^{\infty} n \Pr\{N_B = n|a\} f(a) da \\
&\overset{(a)}{=} \int_a^{\infty} \sum_{n=1}^{\infty} n \frac{(P_m \lambda_{UE} a)^n}{n!} e^{-P_m \lambda_{UE} a} f(a) da \\
&= \int_a^{\infty} P_m \lambda_{UE} a \left(1 - e^{-P_m \lambda_{UE} a}\right) a^{\kappa-1} e^{-\kappa \lambda_{BS} a} \frac{(\kappa \lambda_{BS})^{\kappa}}{\Gamma(\kappa)} da \\
&\overset{(b)}{=} \frac{P_m \lambda_{UE}}{\kappa \lambda_{BS}} \frac{\Gamma(\kappa+1)}{\Gamma(\kappa)} \left(1 - \left(\frac{\kappa \lambda_{BS}}{\kappa \lambda_{BS} + P_m \lambda_{UE}}\right)^{\kappa+1}\right) \\
&= \frac{P_m \lambda_{UE}}{\lambda_{BS}} \left(1 - \frac{1}{\left(1 + \frac{P_m \lambda_{UE}}{\kappa \lambda_{BS}}\right)^{\kappa+1}}\right).
\end{aligned}
\tag{5.18}
$$

Here, (a) holds due to the fact that N_B is a random variable following a Poisson distribution with a mean value $P_m \lambda_{UE} a$ [42]. (b) is derived based on the definition of the gamma function

$$\Gamma(z) = \int_0^{\infty} x^{z-1} e^{-x} dx.$$

Hence, Lemma 5.1 is proved.   □

5.5.2 Nearest SBS Transmission Rate Analysis

In case of content miss, user requested contents are retrieved from the remote servers via backhaul links. Before this, the user first associates to the nearest SBS which provides the highest transmission rate. In this subsection, we provide the analysis of the transmission rate of this cellular link.

Recall that the overlay scheme is employed in the system, and a total of $(1-\phi)W$ bandwidth is allocated to cellular communications. Each SBS serves its associated users in a TDMA manner. Those associated users can be divided into two groups: content miss users whose requested content is not cached locally and users whose requested contents are cached in the SBS cluster. Thus, the users associated to each SBS can be modeled as a PPP Φ_C with parameter $\lambda_C = (P_m + P_s)\lambda_{UE}$. Denote by N_{BS} the total number of SBS. Similar to (5.18), the average cell load can be obtained by

$$\mathbb{E}[N_C] = \frac{(P_m + P_s)\lambda_{UE}}{\lambda_{BS}} \left(1 - \frac{1}{\left(1 + \frac{(P_m+P_s)\lambda_{UE}}{\kappa\lambda_{BS}} \right)^{\kappa+1}} \right). \tag{5.19}$$

Lemma 5.2 *If a user associates to its nearest SBS $\mathcal{B}_o$, a lower bound of the average transmission rate is given by*

$$\mathbb{E}[R_N] \geq \frac{(1-\phi)W}{\mathbb{E}[N_C]\ln 2} \left(2\ln\frac{G_m}{\bar{G}} + \frac{(\alpha-2)\gamma}{2} - \ln J_1(\alpha) \right) \tag{5.20}$$

where

$$J_1(\alpha) = \begin{cases} \dfrac{\Gamma(N_{BS} + 1 - \frac{\alpha}{2})}{(1-\frac{\alpha}{2})\Gamma(N_{BS})} - \Gamma\left(1 - \dfrac{\alpha}{2}\right), & \alpha \neq 2 \\[2mm] \ln(N_{BS} - 1) + \gamma, & \alpha = 2 \end{cases} \tag{5.21}$$

and γ represents the Euler-Mascheroni constant, which equals to 0.577 approximately.

Proof Consider the directional antenna case, in which antenna beams of the user and the associated SBS are aligned, and the received signal strength is given by

$$S = P_B G_m^2 h_1 C r_1^{-\alpha} \tag{5.22}$$

where r_1 denotes the distance from the user to the associated SBS. Note that the interference consists of the signal from all the other SBSs; each of those interference signals is an independent random variable [43].

Based on the transmission model in (5.6), we can obtain the average transmission rate as

$$
\begin{aligned}
\mathbb{E}[R_N] &= \mathbb{E}\left[\frac{(1-\phi)W}{N_C}\log_2\left(1+\frac{S}{I+\sigma^2}\right)\right] \\
&= \frac{(1-\phi)W}{\mathbb{E}[N_C]\ln 2}\mathbb{E}\left[\ln\left(1+\frac{S}{\sum_{i\in\Phi_{BS}\setminus\mathcal{B}_o}I_i+\sigma^2}\right)\right] \\
&\geq \frac{(1-\phi)W}{\mathbb{E}[N_C]\ln 2}\mathbb{E}\left[\ln\frac{S}{\sum_{i\in\Phi_{BS}\setminus\mathcal{B}_o}I_i}\right] \\
&\geq \frac{(1-\phi)W}{\mathbb{E}[N_C]\ln 2}\left(\mathbb{E}[\ln S]-\ln\sum_{i\in\Phi_{BS}\setminus\mathcal{B}_o}\mathbb{E}[I_i]\right).
\end{aligned}
\tag{5.23}
$$

The second equation holds due to the fact that the received signal-to-interference-plus-noise-ratio (SINR) and the cellular load are independent random variables. The first inequality holds because the thermal noise is negligible in high SNR scenarios. In practice, high SNR should be guaranteed due to the reliable communication requirements. The last inequality is directly derived from the Jensen inequality. In what follows, we analyze $\mathbb{E}[\ln S]$ and $\sum_{i\in\Phi_{BS}\setminus\mathcal{B}_o}\mathbb{E}[I_i]$, respectively.

Firstly, from (5.22), the explicit form of $\mathbb{E}[\ln S]$ can be given as

$$
\begin{aligned}
\mathbb{E}[\ln S] &= \mathbb{E}\left[\ln\left(P_B G_m^2 h_1 C r_1^{-\alpha}\right)\right] \\
&= \ln\left(P_B G_m^2 C\right) + \mathbb{E}[\ln h_1] - \alpha\mathbb{E}[\ln r_1] \\
&= \ln\left(P_B G_m^2 C\right) - \gamma + \frac{\alpha}{2}\left(\gamma + \ln\pi\lambda_{BS}\right).
\end{aligned}
\tag{5.24}
$$

The equality holds due to the fact that

$$
\mathbb{E}[\ln h_1] = \int_0^\infty \ln x e^{-x}dx = -\gamma
$$

and

$$
\begin{aligned}
\mathbb{E}[\ln r_1] &= \int_0^\infty \ln r_1 f(r_1)dr_1 \\
&\overset{(a)}{=} \int_0^\infty \ln(r_1)2\pi\lambda_{BS}r_1 e^{-\pi\lambda_{BS}r_1^2}dr_1 \\
&\overset{(b)}{=} \frac{1}{2}\left(\int_0^\infty e^{-y}\ln y\,dy - \int_0^\infty e^{-y}\ln(\pi\lambda_{BS})dy\right) \\
&= -\frac{\gamma+\ln\pi\lambda_{BS}}{2}
\end{aligned}
\tag{5.25}
$$

where (a) holds because r_1 follows distribution [43]

$$f(r_1) = 2\pi \lambda_{BS} r_1 e^{-\pi \lambda_{BS} r_1^2}.$$

Here, (b) holds by defining an additional variable $y = \pi \lambda_{BS} r_1^2$.

Secondly, based on the average interference model in (5.9), the value of $\sum_{i \in \Phi_{BS} \setminus \mathcal{B}_o} \mathbb{E}[I_i]$ is given as

$$
\begin{aligned}
\ln \sum_{i \in \Phi_{BS} \setminus \mathcal{B}_o} \mathbb{E}[I_i] &= \ln \sum_{i \in \Phi_{BS} \setminus \mathcal{B}_o} P_B \bar{G}^2 C \mathbb{E}[h_i] \mathbb{E}[r_i^{-\alpha}] \\
&= \ln \left(P_B \bar{G}^2 C \right) + \ln \sum_{i \in \Phi_{BS} \setminus \mathcal{B}_o} \mathbb{E}[r_i^{-\alpha}].
\end{aligned}
\tag{5.26}
$$

The last equality holds because $\mathbb{E}[h_i] = 1$.

Recall that the geographic locations of both users and SBSs follow PPP; the PDF of the distance between a user and its i-th nearest SBS r_i can be expressed as [42]

$$f(r, i) = \frac{2(\pi \lambda_{BS})^i}{(i-1)!} r^{2i-1} e^{-\pi \lambda_{BS} r^2}, \; \forall i = 2, 3 \dots$$

then, the $-\alpha$th moment of r_i can be calculated as follows:

$$
\begin{aligned}
\mathbb{E}[r_i^{-\alpha}] &= \int_0^\infty r^{-\alpha} f(r, i) dr \\
&= \int_0^\infty \frac{2(\pi \lambda_{BS})^i}{(i-1)!} r^{2i-1-\alpha} e^{-\pi \lambda_{BS} r^2} dr \\
&= \frac{(\pi \lambda_{BS})^{\frac{\alpha}{2}}}{(i-1)!} \int_0^\infty y^{\frac{2i-\alpha}{2}-1} e^{-y} dy \\
&= (\pi \lambda_{BS})^{\frac{\alpha}{2}} \frac{\Gamma(i - \frac{\alpha}{2})}{\Gamma(i)}, \text{ if } i > \frac{\alpha}{2}.
\end{aligned}
\tag{5.27}
$$

Now, we turn to the summation of the $-\alpha$th moments of r_i when i varies.

- When $\alpha \neq 2$, from (5.27), the summation of the $-\alpha$th moments of r_i, for $i = 2, 3, \cdots$, is given by

$$
\begin{aligned}
\sum_{i \in \Phi_{BS} \setminus \mathcal{B}_o} \mathbb{E}[r_i^{-\alpha}] &= (\pi \lambda_{BS})^{\frac{\alpha}{2}} \sum_{i=2}^{N_{BS}} \frac{\Gamma(i - \frac{\alpha}{2})}{\Gamma(i)} \\
&= (\pi \lambda_{BS})^{\frac{\alpha}{2}} \left(\frac{\Gamma(N_{BS} + 1 - \frac{\alpha}{2})}{(1 - \frac{\alpha}{2}) \Gamma(N_{BS})} - \Gamma \left(1 - \frac{\alpha}{2} \right) \right).
\end{aligned}
\tag{5.28}
$$

The last equality is a direct application of the following equality [43]:

$$\sum_{j=1}^{n} \frac{\Gamma(j-\beta)}{\Gamma(j)} = \frac{\Gamma(n+1-\beta)}{(1-\beta)\Gamma(n)}, \beta \neq 2. \tag{5.29}$$

- When $\alpha = 2$, the summation of the moments is given by

$$\sum_{i \in \Phi_{BS} \setminus \mathcal{B}_o} \mathbb{E}[r_i^{-\alpha}] = (\pi\lambda_{BS})^{\frac{\alpha}{2}} \sum_{i=2}^{N_{BS}} \frac{\Gamma(i-1)}{\Gamma(i)}$$

$$= (\pi\lambda_{BS})^{\frac{\alpha}{2}} \sum_{i=1}^{N_{BS}-1} \frac{1}{i} \tag{5.30}$$

$$\approx (\pi\lambda_{BS})^{\frac{\alpha}{2}} \left(\ln(N_{BS}-1) + \gamma\right)$$

where the equality holds if N_{BS} is sufficiently large, which is true in dense networks.

By combining (5.28) and (5.30) with (5.26), the summation of average interference power can be obtained in this logarithmic form as

$$\ln \sum_{i \in \Phi_{BS} \setminus \mathcal{B}_o} \mathbb{E}[I_i] = \ln P_B \bar{G}^2 C + \frac{\alpha}{2} \ln \pi\lambda_{BS} + \ln J_1(\alpha) \tag{5.31}$$

where $J_1(\alpha)$ is defined in (5.21). Substituting (5.24) and (5.31) into (5.23) concludes the proof of Lemma 5.2. $\qquad\qquad\square$

Remark 5.1 Lemma 5.2 characterizes the impact of system parameters on the performance of the nearest SBS transmission, including the network density, the directional antenna gain, and the path loss exponent. We can have the following observations:

- Firstly, the average transmission rate grows linearly with the directional antennas gain, i.e., $G_m/\bar{G}$, indicating that the employment of directional antennas can effectively improve the throughput of mmWave systems.
- Secondly, the average transmission rate increases linearly with the path loss exponent α; this is because the high path loss of mmWave signals largely mitigates mutual interference and contributes to a spatial reuse gain.
- Finally, the transmission performance slightly drops with network density, since the value of $\ln J_1(\alpha)$ slightly increases with the number of SBSs N_{BS}. This is because the distance of both communication signals and interference signals is affected by the network density.

5.5.3 SBS Cluster Transmission Rate Analysis

Recall that each SBS in the cluster is designed to cache contents with equal overall content hit ratio; each user thus associates to any SBS in the cluster with the same probability, which means that the average transmission rate should be averaged over all the candidate SBSs.

Lemma 5.3 *The average transmission rate of users associating to SBS clusters is given by*

$$
\mathbb{E}[R_C] \geq \frac{(1-\phi)W}{\mathbb{E}[N_C]\ln 2} \left(2\ln \frac{G_m}{\bar{G}} + \frac{(\alpha-2)\gamma}{2} \right.
$$
$$
\left. - \frac{\alpha}{2K} \sum_{k=1}^{K} \sum_{i=1}^{k-1} \frac{1}{i} - \frac{1}{K} \sum_{k=1}^{K} \ln J_2(\alpha, k) \right)
$$
(5.32)

where

$$
J_2(\alpha, k) = \begin{cases} \dfrac{\Gamma(N_{BS}+1-\frac{\alpha}{2})}{(1-\frac{\alpha}{2})\Gamma(N_{BS})} - \dfrac{\Gamma(k-\frac{\alpha}{2})}{\Gamma(k)}, & \alpha < 2 \\[2ex] E_1(r_0) + \ln(N_{BS}-1) + \gamma - J_4(k), & \alpha = 2 \\[2ex] \Gamma\left(1-\dfrac{\alpha}{2}, r_0\right) + \dfrac{\Gamma(N_{BS}+1-\frac{\alpha}{2})}{(1-\frac{\alpha}{2})\Gamma(N_{BS})} - \Gamma\left(1-\dfrac{\alpha}{2}\right) - J_3(k) \end{cases}
$$
(5.33)

$$
J_3(k) = \begin{cases} \Gamma\left(1-\dfrac{\alpha}{2}, r_0\right), & k = 1 \\[2ex] \dfrac{\Gamma\left(k-\frac{\alpha}{2}\right)}{\Gamma(k)}, & k \geq 2, \end{cases}
$$
(5.34)

$$
J_4(k) = \begin{cases} E_1(r_0), & k = 1 \\[2ex] \dfrac{1}{k-1}, & k \geq 2. \end{cases}
$$
(5.35)

where $r_0 = \pi \lambda_{BS} d_0^2$. *Here,*

$$
\Gamma(z, a) = \int_a^\infty x^{z-1} e^{-x} dx
$$

and

$$E_1(x) = \int_x^\infty \frac{1}{t} e^{-t} dt$$

are the incomplete gamma function and the exponential integral function, respectively.

Proof Denote by $\mathcal{B}_o = \{\mathcal{B}_o^1, \mathcal{B}_o^2, \ldots, \mathcal{B}_o^k, \ldots, \mathcal{B}_o^K\}$ the set of candidate SBSs from the cluster, sorted according to the physical distances in ascending order. The corresponding set of physical distances is $\{r_1, r_2, \ldots, r_k, \ldots, r_K\}$.

If a user associates to the kth nearest SBS $\mathcal{B}_o^k$, the expected received signal strength is given by

$$S_C^k = P_B G_m^2 h_1 C r_k^{-\alpha}, 1 \le k \le K. \tag{5.36}$$

The corresponding interference strength is given by

$$I_C^k = \sum_{i \in \Phi_{BS} \setminus \mathcal{B}_o^k} P_B G(\theta_{t,i}) G(\theta_{r,i}) h_i C r_i^{-\alpha}, 1 \le k \le K. \tag{5.37}$$

Let R_C^k denote the average transmission rate between the user and its kth nearest SBS; the average transmission rate among all the SBSs in the cluster is given by

$$
\begin{aligned}
\mathbb{E}[R_C] &= \mathbb{E}\left[\frac{1}{K} \sum_{k=1}^K R_C^k \right] \\
&= \frac{(1-\phi)W}{K \mathbb{E}[N_C] \ln 2} \sum_{k=1}^K \ln\left(1 + \frac{S_C^k}{I_C^k + \sigma^2} \right) \\
&\ge \frac{(1-\phi)W}{K \mathbb{E}[N_C] \ln 2} \sum_{k=1}^K \left(\mathbb{E}[\ln S_C^k] - \ln \sum_{i \in \Phi_{BS} \setminus \mathcal{B}_o^k} I_i^k \right) \\
&= \frac{(1-\phi)W}{\mathbb{E}[N_C] \ln 2} \left(2 \ln\left(\frac{G_m}{\bar{G}} \right) - \gamma - \frac{\alpha}{K} \sum_{k=1}^K \mathbb{E}[\ln r_k] \right. \\
&\quad \left. - \frac{1}{K} \sum_{k=1}^K \ln \sum_{i \in \Phi_{BS} \setminus \mathcal{B}_o^k} \mathbb{E}[r_i^{-\alpha}] \right).
\end{aligned}
\tag{5.38}
$$

The inequality holds due to the lower bound in (5.23). The last step holds due to (5.24) and (5.26). Then, according to [25], $\mathbb{E}[\ln r_k]$ is given by

$$\mathbb{E}[\ln r_k] = -\frac{1}{2}\left(\gamma + \ln(\pi\lambda_{BS}) - \sum_{i=1}^{k-1}\frac{1}{i}\right). \tag{5.39}$$

Then, we derive the expression of $\sum_{i\in\Phi_{BS}\setminus\mathcal{B}_o^k}\mathbb{E}[r_i^{-\alpha}]$ in different cases, respectively.

- When $\alpha < 2$, from (5.27), we have

$$\sum_{i\in\Phi_{BS}\setminus\mathcal{B}_o^k}\mathbb{E}[r_i^{-\alpha}] = \sum_{i=1}^{N_{BS}}\mathbb{E}[r_i^{-\alpha}] - \mathbb{E}[r_k^{-\alpha}]$$

$$= (\pi\lambda_{BS})^{\frac{\alpha}{2}}\left(\frac{\Gamma(N_{BS}+1-\frac{\alpha}{2})}{(1-\frac{\alpha}{2})\Gamma(N_{BS})} - \frac{\Gamma(k-\frac{\alpha}{2})}{\Gamma(k)}\right). \tag{5.40}$$

- When $\alpha = 2$, $\mathbb{E}[r_1^{-\alpha}] = \int_{r_0}^{\infty}\frac{1}{y}e^{-y}dy = E_1(r_0)$ according to the exponential integral function $E_1(x)$. From (5.30), if $k \geq 2$, it holds that $\mathbb{E}[r_k^{-\alpha}] = \frac{1}{k-1}$. Hence, $\mathbb{E}[r_k^{-\alpha}]$ can be represented by a piecewise function $J_4(k)$ given in (5.35). Similar to (5.43), we have

$$\sum_{i\in\Phi_{BS}\setminus\mathcal{B}_o^k}\mathbb{E}[r_i^{-\alpha}] = (\pi\lambda_{BS})\left(E_1(r_0) + \ln(N_{BS}-1) + \gamma - J_4(k)\right). \tag{5.41}$$

- When $\alpha > 2$, $\mathbb{E}[r_1^{-\alpha}]$ cannot be bounded as the condition $i > \frac{\alpha}{2}$ in (5.27) is no longer met. This is because the path loss model in (5.4) becomes invalid when the distance is less than d_0. To address this issue, a guard radius d_0 is imposed to receivers in order to exclude interference origins in the short distance. Hence, $-\alpha$th moments of r_1 become

$$\mathbb{E}[r_1^{-\alpha}] = \int_{d_0}^{\infty} r_1^{-\alpha} f(r_1)dr_1$$

$$= (\pi\lambda_{BS})^{\frac{\alpha}{2}}\int_{\pi\lambda_{BS}d_0^2}^{\infty} y^{\frac{2-\alpha}{2}-1}e^{-y}dy \tag{5.42}$$

$$= (\pi\lambda_{BS})^{\frac{\alpha}{2}}\Gamma\left(1-\frac{\alpha}{2}, r_0\right).$$

Thus, $\sum_{i\in\Phi_{BS}\setminus\mathcal{B}_o^k}\mathbb{E}[r_i^{-\alpha}]$ is given by

$$\sum_{i \in \Phi_{BS} \setminus \mathcal{B}_o^k} \mathbb{E}[r_i^{-\alpha}] = \mathbb{E}[r_1^{-\alpha}] + \sum_{i=2}^{N_{BS}} \mathbb{E}[r_i^{-\alpha}] - \mathbb{E}[r_k^{-\alpha}]$$

$$= (\pi \lambda_{BS})^{\frac{\alpha}{2}} \left(\Gamma\left(1 - \frac{\alpha}{2}, r_0\right) - \Gamma\left(1 - \frac{\alpha}{2}\right) \right. \tag{5.43}$$

$$\left. + \frac{\Gamma\left(N_{BS} + 1 - \frac{\alpha}{2}\right)}{\left(1 - \frac{\alpha}{2}\right)\Gamma(N_{BS})} - J_3(k) \right).$$

By substituting (5.39), (5.40), (5.40) and (5.41) into (5.38) concludes the proof of Lemma 5.3.   □

Remark 5.2 Lemma 5.3 characterizes the transmission performance of the SBS cluster with respect to different system parameters. Here are some important observations. Similar to the conclusion in Lemma 5.2, the average SBS cluster transmission rate slightly drops with network density, since $J_2(\alpha, k)$ slightly grows with network density. This lemma also shows that the transmission rate reduces with the cluster size, which highlights the balance between caching diversity and transmission efficiency. Increasing cluster size leads to a higher cache capacity that can cache more contents, yet, the transmission performance drops as users are required to retrieve contents from a longer distance.

5.5.4 D2D Transmission Rate Analysis

In this subsection, we analyze the D2D caching performance. Recall that the distribution of D2D users follows a homogeneous PPP, Φ_D, with density $\lambda_D = P_d \lambda_{UE}$. Compared with SBS, mobile users obtain a smaller directional antenna gain, due to limited antenna space. Let G_u^m and $\bar{G}_u$ represent the maximal and average antenna gain of users' main lobe, respectively. In an overlay scheme, a total of ϕW system bandwidth is allocated to D2D communications.

Lemma 5.4 *A lower bound of the average D2D transmission rate can be given by*

$$\mathbb{E}[R_D] \geq \frac{\phi W}{\ln 2} \left(2 \ln \frac{G_u^m}{\bar{G}_u} - \gamma - \alpha \left(\ln r_d^{max} - \frac{1}{2} \right) - \ln(\pi \lambda_D) - \ln J_5(\alpha) \right) \tag{5.44}$$

where

$$J_5(\alpha) = \begin{cases} \dfrac{R^{2-\alpha}}{1 - \frac{\alpha}{2}}, & \alpha < 2 \\[2ex] 2\ln\left(\dfrac{R}{d_0}\right), & \alpha = 2 \\[2ex] \dfrac{R^{2-\alpha} - d_0^{2-\alpha}}{1 - \frac{\alpha}{2}}, & \alpha > 2 \end{cases} \tag{5.45}$$

where $R = \sqrt{\dfrac{N_D}{\pi \lambda_D}}$ and N_D is the amount of D2D transmitters.

Proof If a user retrieves contents from its D2D peer via a mmWave link, the received signal power is determined by

$$S_D = P_U (G_u^m)^2 h_1 C r_d^{-\alpha}. \tag{5.46}$$

The received interference, which consists of signals from all the other D2D transmitters Φ_D, is expressed as

$$I_D = \sum_{i \in \Phi_D} P_U G_u(\theta_{t,i}) G_u(\theta_{r,i}) h_i C r_i^{-\alpha} \tag{5.47}$$

where r_i is the distance from the user to the i-th source of D2D interference.

Similarly, a lower bound of the average D2D transmission rate can be obtained as follows:

$$\begin{aligned} \mathbb{E}[R_D] &= \phi W \log_2\left(1 + \frac{S_D}{I_D + \sigma^2}\right) \\[1ex] &\geq \frac{\phi W}{\ln 2}\left(\mathbb{E}[\ln S_D] - \ln \sum_{i \in \Phi_D} \mathbb{E}[I_D^i]\right) \\[1ex] &\geq \frac{\phi W}{\ln 2}\left(2\ln \frac{G_u^m}{\bar{G}_u} - \gamma - \alpha \mathbb{E}[\ln r_d] - \ln \sum_{i \in \Phi_D} \mathbb{E}[r_i^{-\alpha}]\right). \end{aligned} \tag{5.48}$$

Using the same approach in (5.24), combining the PDF of r_d in (5.1), $\mathbb{E}[\ln r_d]$ is given by

$$\begin{aligned} \mathbb{E}[\ln r_d] &= \int_0^{r_d^{max}} \ln r_d \frac{2r_d}{(r_d^{max})^2} dr_d \\[1ex] &= \ln r_d^{max} - \frac{1}{2}. \end{aligned} \tag{5.49}$$

For the term $\sum_{i\in\Phi_D} \mathbb{E}[r_i^{-\alpha}]$ in (5.48), it can be obtained in the following way. Note that r_i denotes the inter-node distances of PPP, which follows a generalized beta distribution [43]. The $-\alpha$th moment of r_i is given by

$$
\mathbb{E}[r_i^{-\alpha}] =
\begin{cases}
\dfrac{R^{-\alpha}\Gamma(N_D+1)\Gamma(i-\frac{\alpha}{2})}{\Gamma(i)\Gamma(N_D+1-\frac{\alpha}{2})}, & \alpha < 2 \\[4mm]
\infty, & \alpha \geq 2
\end{cases}
\tag{5.50}
$$

where $R = \sqrt{\frac{N_D}{\pi\lambda_D}}$ is the equivalent cell radius and N_D is the number of D2D users. The summation of the $-\alpha$th moments of r_i can be generalized into different forms when the value of α varies.

- When $\alpha < 2$, the summation of moment can be obtained by

$$
\sum_{i\in\Phi_D} \mathbb{E}[r_i^{-\alpha}] = \frac{R^{-\alpha}\Gamma(N_D+1)}{\Gamma(N_D+1-\frac{\alpha}{2})} \sum_{i=1}^{N_D} \frac{\Gamma(i-\frac{\alpha}{2})}{\Gamma(i)}
$$

$$
= \pi\lambda_D \frac{R^{2-\alpha}}{1-\frac{\alpha}{2}}
\tag{5.51}
$$

where the last equality holds due to the fact that $\pi R^2\lambda_D = N_D$.
- When $\alpha > 2$, the moment of r_i cannot be bounded as the adopted path loss model no longer holds at short distances. This case can be tackled by imposing a guard radius d_0 around each receiver, i.e., data transmission with a distance less than d_0 is not allowed. According to (5.52), the sum of interference within distance d_0 is expressed as $\pi\lambda_D d_0^{2-\alpha}/(1-\frac{\alpha}{2})$. By excluding the interference inside the guard radius, we have

$$
\sum_{i\in\Phi_D} \mathbb{E}[r_i^{-\alpha}] = \pi\lambda_D \frac{R^{2-\alpha} - d_0^{2-\alpha}}{1-\frac{\alpha}{2}}.
\tag{5.52}
$$

- When $\alpha = 2$, taking limits on the right-hand side of (5.51), we have

$$
\sum_{i\in\Phi_D} \mathbb{E}[r_i^{-\alpha}] = 2\pi\lambda_D \ln\left(\frac{R}{d_0}\right).
\tag{5.53}
$$

By substituting (5.49), (5.51), (5.51), and (5.53) into (5.48), Lemma 5.4 is proved.   □

Remark 5.3 Lemma 5.4 characterizes the impact of physical layer parameters on the transmission performance of D2D communications. Most notably, it is shown by analytical results that the transmission performance reduces as the user density grows. This is due to the fact that the transmission distance of interference links scales with the density of network, while that of D2D link remains unchanged. With Lemmas 5.2, 5.3, and 5.4, we may observe that the average D2D transmission rate decreases dramatically with the network density, while the performance of cellular transmissions is less sensitive to network density. Consequently, the content retrieval delay via D2D communications highly depends on the network density, necessitating a coordination scheme for D2D communications.

5.6 Performance Evaluation

In this section, the analytical results are validated via extensive Monte-Carlo simulations. Meanwhile, we evaluate the proposed DCEC policy by comparing with the state-of-the-art benchmark polices. Firstly, the simulation setup is given in Sect. 5.6.1, followed by the backhaul offloading performance evaluation in Sect. 5.6.2. Analytical results on transmission performance and content retrieval delay are presented in Sects. 5.6.3 and 5.6.4, respectively.

5.6.1 Simulation Setup

Table 5.2 summarizes important simulation parameters. We simulate a plane network with area 1 km^2 (1000 m × 1000 m). For the mmWave system, we consider the ratified IEEE 802.11ad standard operating at the 60 GHz unlicensed band. A total of 2.16 GHz bandwidth is allocated [44], in which 80% of the bandwidth is allocated to cellular communications and 20% to D2D communications. The model parameters of the directional antennas are configured empirically [33]. Totally, we consider three typical scenarios: a conference room with LOS connections and a living room with LOS links and with NLOS links. Their corresponding path loss exponents are set to be 1.4, 1.6, and 2, respectively. Due to the limited battery and space on mobile devices, mobile users can only support a lower transmitting power and less directional antenna gain, as compared with SBS. The backhaul capacity is constrained and set to be 3 Gbit/s unless otherwise specified. The SBS density of the mmWave networks ranges from 80 to 400 per km^2, with an average cell radius varying from 65 to 30 m, corresponding to the scope from sparse rural to dense urban. The user density is configured from 800 to 4000 per km^2, where 80% of the users are successfully paired. We consider a library of 2000 contents in total. The cache capacity of users and SBSs is set to be 150 and 200 (in unit of contents), respectively. Unless specified, the application of video streaming is considered, which has a content popularity skewness of 0.56 [26].

The state-of-the-art most popular caching (MPC) policy is adopted as the comparison benchmark. In this policy, the user and its associated SBS only collectively cache the most popular contents, i.e., $\{f_1, f_2, \ldots, f_{C_u}\}$ and $\{f_{C_u+1}, f_{C_u+2}, \ldots, f_{C_u+C_s}\}$, respectively.

Table 5.2 Simulation parameters in cooperative caching

Notation	Parameter	Value
A	Simulation area	1 km^2
N_o	Background noise density	-174 dBm/Hz
W	Bandwidth	2.16 GHz
ϕ	Fraction of D2D spectrum	20%
f	Carrier frequency	60 GHz
α	Path loss exponent	{1.4, 1.6, 2}
P_B	SBS transmit power	30 dBm
P_U	User transmit power	20 dBm
G_s^m	SBS main lobe gain	18 dB
G_s^s	SBS side lobe gain	-2 dB
G_u^m	User main lobe gain	9 dB
G_u^m	User side lobe gain	-2 dB
ω_m	Half-power beamwidth	10°
d_0	Reference distance	1 m
r_d^{max}	Maximum D2D distance	10 m
λ_{BS}	Network density	{80–400} per km^2
λ_{UE}	User density	{800–4000} per km^2
δ	Fraction of paired users	80%
F	Content library size	2000
ν	Average content size	100 Mbit
C_u	User cache capacity	150
C_s	SBS cache capacity	200

5.6.2 Backhaul Offloading Performance

Figure 5.2 shows the comparison of backhaul offloading performance with varying content popularity skewness. Clearly, the proposed DCEC policy significantly outperforms the benchmark policy, since it is able to exploit the cache resource in an efficient manner. In specific, the DCEC policy can offload about more than 50% of the backhaul traffic than the MPC policy when content skewness $\xi = 0.6$. It is observed that the gap narrows down with the increase of popularity skewness. This is because the cache capacities of an individual user and its associated SBS are adequate to store contents with highly concentrated popularity.

Figure 5.3 demonstrates how the SBS cache capacity affects the backhaul offloading performance. Clearly, the increase of SBS cache capacity leads to a higher portion of offloaded backhaul traffic, as more contents can be cached locally. Interestingly, the caching capacity growth provides a higher gain in the small-skewness region than that in the large-skewness region. This is because the higher the skewness, the more concentrated the content popularity, and hence less cache resources can efficiently provisioning more content requests.

We further validate the performance of backhaul offloading with respect to the SBS cluster size, as shown in Fig. 5.4, significant backhaul offloading gain can be

Fig. 5.2 Backhaul offloading performance with respect to content popularity skewness

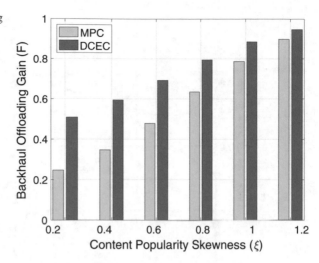

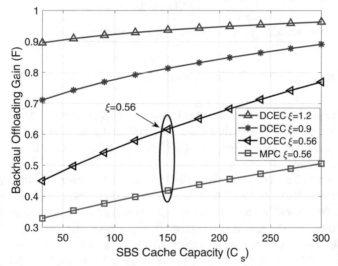

Fig. 5.3 Backhaul offloading performance with respect to SBS cache capacity ($K = 3$)

achieved by increasing the SBS cluster size when the content popularity is less skewed ($\xi = 0.3$ and 0.56). In specific, when $\xi = 0.56$, the proposed DCEC policy with 8 SBSs can offload more than 70% of the backhaul traffic, compared with that of two SBSs. In contrast, for a large value of ξ, the performance gain resulting from a high cluster size becomes marginal. Hence, the proposed DCEC policy is more likely to be applied in applications with less concentrated content popularity.

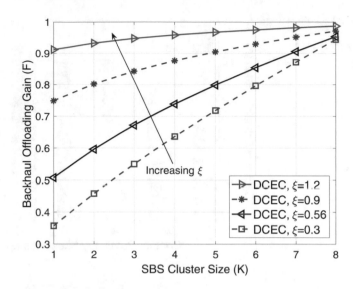

Fig. 5.4 Backhaul offloading performance with respect to SBS cluster size

5.6.3 Transmission Performance

In this subsection, extensive simulations are carried out to validate the analytical results on the transmission performance. To avoid randomness, the presented results are the average over 10,000 samples of different realizations of network topologies and channel fading.

Figure 5.5 shows how the transmission rate of the nearest SBS varies with network density for $\alpha = 1.4, 1.6, 2$. The lower bounds obtained from Lemma 5.2 well match the simulation results under various channel conditions, which confirms the correctness of the analytical results. We can also observe that the average rate slightly drops with network density. In the case of $\alpha = 2$, the reduction of transmission rate is only 8% when network density increases from 80 to 400 per km^2. When $\alpha = 1.4$, the transmission rate reduces by only 15%. Meanwhile, the average transmission rate increases with α because interference is suppressed by severe propagation loss in mmWave channels.

Figure 5.6 shows the average SBS cluster transmission rate in a different scenario of network density, when the value of $\alpha = 1.4, 1.6, 2$. It can be observed that the gap between simulation results and the analytical bounds narrows with network density, which validates the correctness of Lemma 5.3. In addition, the transmission performance is shown to slightly decrease with network density, which is analogous with that of the nearest SBS. In particular, when $\alpha = 2$, the average transmission rate drops by approximately 10%, when the network turns from a sparse one with $\lambda_{BS} = 80$ to a dense one with $\lambda_{BS} = 400$.

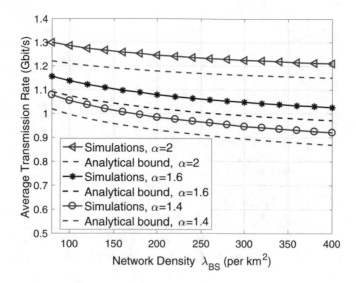

Fig. 5.5 The nearest SBS transmission rate with respect to the network density

Figure 5.7 shows the average transmission rates with respect to the network density at a different cluster size K. Analytical results well match the simulation results, which corroborates the correctness of Lemma 5.3. Notably, the average SBS cluster transmission rate drops with the cluster size. Specifically, the SBS cluster with two SBSs can achieve a data rate of 1.08 Gbit/s when $\lambda_{BS} = 80$, while the SBS cluster with 4 SBSs only attains a data rate of 0.83 Gbit/s at the equal network density, denoting a decrease of nearly 23%. This is due to the fact that users can retrieve their target contents from remote SBSs. A large SBS cluster size indicates that more contents can be achieved while reducing the average transmission rate, which highlights the balance between caching diversity and transmission efficiency.

Figure 5.8 shows the average D2D transmission rate varying with the D2D user density when $\alpha = 1.4, 1.6, 2$. Simulation results are well bounded by the analytical results from Lemma 5.4 in various channel conditions. Firstly, due to short communication distance, D2D communications provide a higher transmission rate than cellular communications, which guarantees a low content retrieval delay. Secondly, a sharp performance degradation is observed when the D2D user density grows, as the transmission rate drops from 4 Gbit/s to only 2 Gbit/s when the D2D user density changes from 40 to 800 per km^2 for $\alpha = 1.6$. This is because the expected signal strength remains the same as D2D communication distance is not affected by user density. In contrast, the interference increases drastically as the distance of interference link reduces with the D2D user density. Consequently, it is required to design a coordinated scheduling policy so as to enhance D2D transmission performance in dense networks.

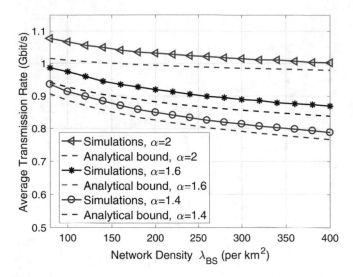

Fig. 5.6 SBS cluster transmission rate with respect to network density ($K = 2$)

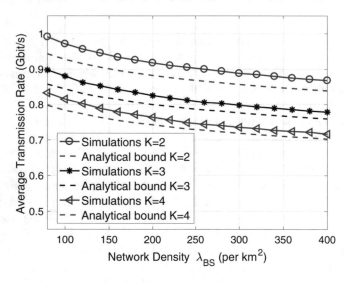

Fig. 5.7 SBS cluster transmission rate with respect to the network density ($\alpha = 1.6$)

5.6.4 Content Retrieval Delay

We evaluate the delay of content retrieval of different caching policies, varying content popularity distributions, ranging network density and backhaul capacity, as well as different cluster sizes.

Figure 5.9 demonstrates how content popularity skewness affects the average delay of content retrieval. It is clear that the proposed DCEC policy outperforms the MPC benchmark in low-popularity skewness region. This is because contents with

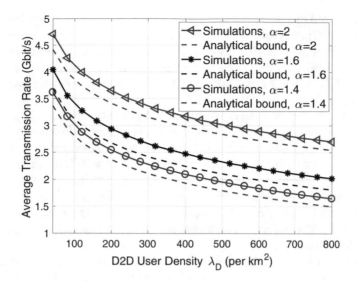

Fig. 5.8 D2D transmission rate with respect to D2D user density

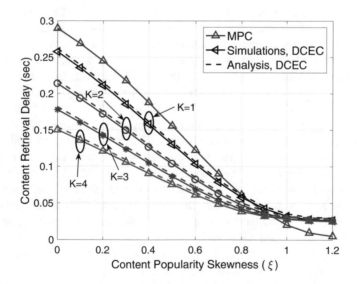

Fig. 5.9 Content retrieval delay with respect to content popularity

less concentrated popularity distribution favor large cache capacity. For instance, the proposed DCEC policy with 4 SBSs can decrease 48% of the content retrieval delay, compared with that of the MPC for $\xi = 0.6$. Nevertheless, the achieved performance gain by the DCEC diminishes with the increase of skewness, which further confirms that the DCEC policy is significant in applications with less concentrated content popularity.

The delay of content retrieval with respect to the network density is evaluated for $\xi = 0.56$. As shown in Fig. 5.10, simulation results well match those of the analytical bounds. Meanwhile, the delay of content retrieval grows with network density, since the transmission rate of both cellular and D2D drops in dense networks. For example, users may spend around 23% more time for content retrieval when network density changes from 80 to 400 per km^2 for $K = 4$. More importantly, with the help of high-rate cellular and D2D communications, the DCEC policy with 4 SBSs is able to reduce as much as 45% of the delay when compared with the MPC benchmark.

As shown in Fig. 5.11, the average transmission delay varies with the backhaul capacity. The performance gain achieved by the DCEC policy becomes marginal with the increase of backhaul capacity. If the backhaul capacity happens to be the bottleneck of the system performance, the proposed DCEC policy cooperatively caches more contents in edge networks, which helps suppress the content retrieval delay. However, if the backhaul capacity is unconstrained, the performance gain drops as users can fetch interested contents at low delay cost from remote servers.

Figure 5.12 demonstrates how the size of SBS cluster affects the average delay of content retrieval for $\lambda_{BS} = 100, 200, 400$. Recall that we should strike a balance between the transmission efficiency and the caching diversity. It is observed that the average delay of content retrieval first drops and then grows with the SBS cluster size, which indicates that if the SBS cluster size is excessively large, the gain of cooperative edge caching diminishes. This is because a large SBS cluster enables the caching of more popular contents. At the meantime, it reduces the average transmission rate as the physical distances for content retrieval become longer. Lastly, the possible balance between caching diversity and transmission performance implies that there exists an optimal size of SBS cluster. In specific, when $\lambda_{BS} = 100$, the optimal cluster size is 7. If the network density grows to 400 per km^2, the optimal value drops to 6.

Finally, Fig. 5.13 shows the decrease of optimal SBS cluster size with backhaul capacity, which means that a large value of SBS cluster is preferred in backhaul capacity-limited scenarios. This is because higher backhaul capacity can effectively boost the performance gain of the DCEC policy. For instance, the optimal cluster size is 7 when the backhaul capacity is 2 Gbit/s, but it drops to 3 when the capacity becomes 16 Gbit/s.

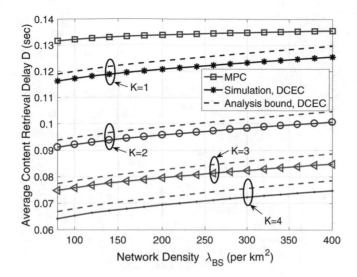

Fig. 5.10 Content retrieval delay with respect to network density

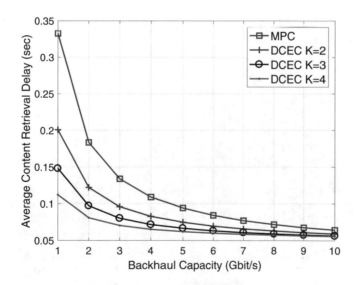

Fig. 5.11 Content retrieval delay with respect to backhaul capacity

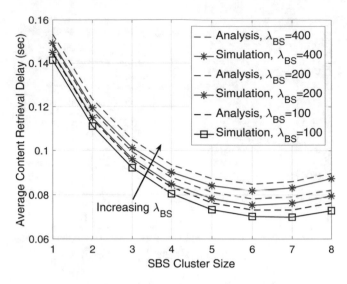

Fig. 5.12 The impact of SBS cluster size on the content retrieval delay

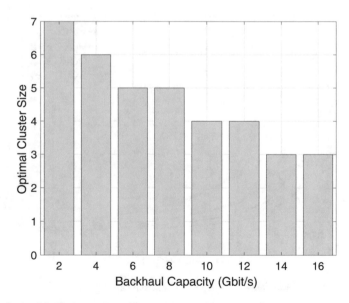

Fig. 5.13 Optimal SBS cluster size with respect to backhaul capacity

5.7 Summary

In this chapter, we have introduced a D2D-assisted cooperative edge caching policy for mmWave dense networks. The closed-form expressions of the backhaul offloading gain and the delay of content retrieval of the introduced policy have been provided, accounting for both the directional antenna model and the network density. It is revealed by both the analytical and simulation results that the average content retrieval delay grows with the network density. In addition, the balance between caching diversity and transmission efficiency has been highlighted. Compared with the state-of-the-art MPC caching benchmark, the introduced policy achieves notable performance gains in terms of backhaul traffic offloading and the delay of content retrieval.

References

1. W. Wu, N. Zhang, N. Cheng, Y. Tang, K. Aldubaikhy, X. Shen, Beef up mmwave dense cellular networks with D2D-assisted cooperative edge caching. IEEE Trans. Veh. Technol. **68**(4), 3890–3904 (2019)
2. M. Xiao, S. Mumtaz, Y. Huang, L. Dai, Y. Li, M. Matthaiou, G.K. Karagiannidis, E. Björnson, K. Yang, I. Chih-Lin, Millimeter wave communications for future mobile networks. IEEE J. Sel. Areas Commun. **35**(9), 1909–1935 (2017)
3. N. Zhang, N. Cheng, A.T. Gamage, K. Zhang, J.W. Mark, X. Shen, Cloud assisted hetnets toward 5G wireless networks. IEEE Commun. Mag. **53**(6), 59–65 (2015)
4. T.S. Rappaport, S. Sun, R. Mayzus, H. Zhao, Y. Azar, K. Wang, G.N. Wong, J.K. Schulz, M. Samimi, F. Gutierrez, Millimeter wave mobile communications for 5G cellular: it will work! IEEE Access **1**, 335–349 (2013)
5. N. Zhang, P. Yang, J. Ren, D. Chen, L. Yu, X. Shen, Synergy of big data and 5G wireless networks: opportunities, approaches, and challenges. IEEE Wireless Commun. **25**(1), 12–18 (2018)
6. E. Bastug, M. Bennis, M. Debbah, Living on the edge: the role of proactive caching in 5G wireless networks. IEEE Commun. Mag. **52**(8), 82–89 (2014)
7. Y. Zhong, M. Haenggi, F. Zheng, W. Zhang, T.Q. Quek, W. Nie, Towards a tractable delay analysis in ultra-dense networks. IEEE Commun. Mag. **55**(12), 103–109 (2017)
8. H. Wu, J. Chen, W. Xu, N. Cheng, W. Shi, L. Wang, X. Shen, Delay-minimized edge caching in heterogeneous vehicular networks: a matching-based approach. IEEE Trans. Wireless Commun. **19**(10), 6409–6424 (2020)
9. N. Zhang, S. Zhang, J. Zheng, X. Fang, J.W. Mark, X. Shen, QoE driven decentralized spectrum sharing in 5G networks: potential game approach. IEEE Trans. Veh. Technol. **66**(9), 7797–7808 (2017)
10. Mobile-edge computing - introductory technical white paper, European Telecommunications Standards Institute, Tech., Rep., Sep. 2014 [Online]. Available: https://portal.etsi.org/. Accessed 27 March 2017
11. Q. Ye, W. Zhuang, L. Li, P. Vigneron, Traffic-load-adaptive medium access control for fully connected mobile ad hoc networks. IEEE Trans. Veh. Technol. **65**(11), 9358–9371 (2016)
12. Y. Chen, N. Zhang, Y. Zhang, X. Chen, W. Wu, X. Shen, Energy efficient dynamic offloading in mobile edge computing for internet of things. IEEE Trans. Cloud Comput. **9**(3), 1050–1060 (2019)

13. T.G. Rodrigues, K. Suto, H. Nishiyama, N. Kato, Hybrid method for minimizing service delay in edge cloud computing through VM migration and transmission power control. IEEE Trans. Comput. **66**(5), 810–819 (2017)

14. T.G. Rodrigues, K. Suto, H. Nishiyama, N. Kato, K. Temma, Cloudlets activation scheme for scalable mobile edge computing with transmission power control and virtual machine migration. IEEE Trans. Comput. **67**(9), 1287–1300 (2018)

15. Y. Zhou, F.R. Yu, J. Chen, Y. Kuo, Resource allocation for information-centric virtualized heterogeneous networks with in-network caching and mobile edge computing. IEEE Trans. Veh. Technol. **66**(12), 11339–11351 (2017)

16. J. Chen, H. Wu, P. Yang, F. Lyu, X. Shen, Cooperative edge caching with location-based and popular contents for vehicular networks. IEEE Trans. Veh. Technol. **69**(9), 10291–10305 (2020)

17. H. Wu, F. Lyu, C. Zhou, J. Chen, L. Wang, X. Shen, Optimal UAV caching and trajectory in aerial-assisted vehicular networks: a learning-based approach. IEEE J. Sel. Areas Commun. **38**(12), 2783–2797 (2020)

18. J. Liu, H. Nishiyama, N. Kato, J. Guo, On the outage probability of device-to-device-communication-enabled multichannel cellular networks: an RSS-threshold-based perspective. IEEE J. Sel. Areas Commun. **34**(1), 163–175 (2016)

19. W. Song, Y. Zhao, W. Zhuang, Stable device pairing for collaborative data dissemination with device-to-device communications. IEEE Int. Things J. **5**(2), 1251–1264 (2018)

20. Z. Zhou, K. Ota, M. Dong, C. Xu, Energy-efficient matching for resource allocation in D2D enabled cellular networks. IEEE Trans. Veh. Technol. **66**(6), 5256–5268 (2017)

21. M. Ji, G. Caire, A.F. Molisch, The throughput-outage tradeoff of wireless one-hop caching networks. IEEE Trans. Inf. Theory **61**(12), 6833–6859 (2015)

22. R. Wang, J. Zhang, S. Song, K.B. Letaief, Mobility-aware caching in D2D networks. IEEE Trans. Wireless Commun. **16**(8), 5001–5015 (2017)

23. N. Zhao, X. Liu, Y. Chen, S. Zhang, Z. Li, B. Chen, M.S. Alouini, Caching D2D connections in small-cell networks. IEEE Trans. Veh. Technol. **67**(10), 12326–12338 (2018)

24. Z. Chen, J. Lee, T.Q. Quek, M. Kountouris, Cooperative caching and transmission design in cluster-centric small cell networks. IEEE Trans. Wireless Commun. **16**(5), 3401–3415 (2017)

25. S. Zhang, P. He, K. Suto, P. Yang, L. Zhao, X. Shen, Cooperative edge caching in user-centric clustered mobile networks. IEEE Trans. Mobile Comput. **17**(8), 1791–1805 (2018)

26. S. Zhang, N. Zhang, P. Yang, X. Shen, Cost-effective cache deployment in mobile heterogeneous networks. IEEE Trans. Veh. Technol. **66**(12), 11264–11276 (2017)

27. J. Xu, K. Ota, M. Dong, Saving energy on the edge: in-memory caching for multi-tier heterogeneous networks. IEEE Commun. Mag. **56**(5), 102–107 (2018)

28. X. Zhao, P. Yuan, H. Li, S. Tang, Collaborative edge caching in context-aware device-to-device networks. IEEE Trans. Veh. Technol. **67**(10), 9583–9596 (2018)

29. P. Yang, N. Zhang, S. Zhang, L. Yu, J. Zhang, X. Shen, Content popularity prediction towards location-aware mobile edge caching. IEEE Trans. Multimedia **21**(4), 915–929 (2019)

30. W. Song, W. Zhuang, Packet assignment under resource constraints with D2D communications. IEEE Network **30**(5), 54–60 (2016)

31. O. Semiari, W. Saad, M. Bennis, B. Maham, Caching meets millimeter wave communications for enhanced mobility management in 5G networks. IEEE Trans. Wireless Commun. **17**(2), 779–793 (2018)

32. M. Ji, G. Caire, A.F. Molisch, Wireless device-to-device caching networks: basic principles and system performance. IEEE J. Sel. Areas Commun. **34**(1), 176–189 (2016)

33. N. Giatsoglou, K. Ntontin, E. Kartsakli, A. Antonopoulos, C. Verikoukis, D2D-aware device caching in mmwave-cellular networks. IEEE J. Sel. Areas Commun. **35**(9), 2025–2037 (2017)

34. Y. Zhong, T.Q. Quek, X. Ge, Heterogeneous cellular networks with spatio-temporal traffic: delay analysis and scheduling. IEEE J. Sel. Areas Commun. **35**(6), 1373–1386 (2017)

35. W. Wu, N. Cheng, N. Zhang, P. Yang, K. Aldubaikhy, X. Shen, Performance analysis and enhancement of beamforming training in 802.11ad. IEEE Trans. Veh. Technol. **69**(5), 5293–5306 (2020)

36. W. Wu, N. Cheng, N. Zhang, P. Yang, W. Zhuang, X. Shen, Fast mmwave beam alignment via correlated bandit learning. IEEE Trans. Wireless Commun. **18**(12), 5894–5908 (2019)
37. S. Singh, R. Mudumbai, U. Madhow, Interference analysis for highly directional 60-GHz mesh networks: the case for rethinking medium access control. IEEE/ACM Trans. Netw. **19**(5), 1513–1527 (2011)
38. G.R. MacCartney, T.S. Rappaport, Rural macrocell path loss models for millimeter wave wireless communications. IEEE J. Sel. Areas Commun. **35**(7), 1663–1677 (2017)
39. W. Wu, Q. Shen, K. Aldubaikhy, N. Cheng, N. Zhang, X. Shen, Enhance the edge with beamforming: performance analysis of beamforming-enabled WLAN, in *Proceedings of the IEEE WiOpt*, 2018
40. T. Bai, R.W. Heath, Coverage and rate analysis for millimeter-wave cellular networks. IEEE Trans. Wireless Commun. **14**(2), 1100–1114 (2015)
41. S.M. Yu, S.L. Kim, Downlink capacity and base station density in cellular networks, in *Proceedings of the IEEE WiOpt* (2013), pp. 119–124
42. M. Haenggi, J.G. Andrews, F. Baccelli, O. Dousse, M. Franceschetti, Stochastic geometry and random graphs for the analysis and design of wireless networks. IEEE J. Sel. Areas Commun. **27**(7), 1029–1046 (2009)
43. S. Srinivasa, M. Haenggi, Distance distributions in finite uniformly random networks: theory and applications. IEEE Trans. Veh. Technol. **59**(2), 940–949 (2010)
44. W. Wu, Q. Shen, M. Wang, X. Shen, Performance analysis of IEEE 802.11.ad downlink hybrid beamforming, in *Proceedings of the IEEE ICC*, 2017

Chapter 6
Summary and Future Directions

6.1 Summary

This monograph provides the latest research work on improving the performance of mmWave networks. Considering the distinct highly directional feature of mmWave communications, it is vital to achieve efficient and, more importantly, practical solutions in mmWave networks. To this end, a comprehensive and systematic study which analyzes and designs the communication layers to address those new challenges is necessary. In the following, we summarize the main contents of the monograph.

6.1.1 Beam Alignment Scheme Design

From the perspective of the physical layer, the antenna of both the transmitter and the receiver should be aligned, in order to establish reliable communication links. The beam alignment (BA) process, however, may incur significant latency at the order of seconds due to the prohibitive complexity of exhaustively searching the entire beam space. Chapter 3 presents an efficient BA algorithm for mmWave communications. Specifically, the BA process is formulated as a multi-armed bandit (MAB) problem. A learning-based algorithm, named HBA, has been proposed to solve the MAB problem. The HBA algorithm leverages the correlation structure among beams and the prior knowledge on the channel fluctuation to speed up the BA process. Theoretical analysis has been provided to characterize the asymptotic optimality and convergence speed of the HBA algorithm. Through extensive simulations, the effectiveness of the proposed algorithm has been validated, as compared to existing BA method in 802.11ad.

© The Author(s), under exclusive license to Springer Nature Switzerland AG 2021
P. Yang et al., *Millimeter-Wave Networks*, Wireless Networks,
https://doi.org/10.1007/978-3-030-88630-1_6

6.1.2 MAC Performance Evaluation and Enhancement

In medium access control (MAC) layer, in order to coordinate the beamforming (BF) training among multiple users, the IEEE 802.11ad protocol requires all the users to compete for BF training resources without any coordination, which results in frequent collisions in the BF training stage. Thus, how to optimize the MAC parameters toward the maximal global welfare of the system is critical to a mmWave network. In order to effectively manage the network, an accurate modeling of the MAC performance is of paramount importance. To this end, Chap. 4 is dedicated to the evaluation of the MAC performance of mmWave networks. We have focused on the BFT-MAC specified by 802.11ad and have established a simple yet accurate analytical model to evaluate the performance of BFT-MAC. Based on the analytical model, the normalized throughput and average BF training latency have been derived, accounting for different user densities and configurations of BFT-MAC. The derived analytical model provides insightful guidance on practical configurations of BFT-MAC in different scenarios. To improve the performance of BFT-MAC in high user density scenarios, we have proposed an enhancement scheme which adaptively configures the MAC parameters based on user density. Extensive simulations have been performed, and the results validate the correctness of the derived analytical model and the effectiveness of the proposed enhancement scheme.

6.1.3 Backhaul Alleviation Scheme Design

From the perspective of the network layer, it is infeasible to deploy unconstrained wired backhaul links in mmWave dense networks due to prohibitive costs, which results in backhaul congestion, especially in urban scenarios. Thus, developing effective solutions to alleviate the backhaul burden links is essential. In this regard, we resort to the emerging edge caching technology which proactively stores popular contents in user's proximity during off-peak hours. In Chap. 5, we target at developing efficient caching policy to alleviate the backhaul congestion and reduce content retrieval delay in mmWave dense networks. We have proposed the DCEC policy by leveraging the caching resources of both mobile users and small base stations (SBSs) to increase the set of cache contents. Considering both the directional antenna and network density, we have applied the theory of stochastic geometry to derive closed-form expressions on the content retrieval delay performance of the proposed caching policy. Analytical results provide practical guidelines to future mmWave deployment. Comprehensive simulation results validate the accuracy of our analytical results and demonstrate that the proposed caching policy can effectively alleviate the backhaul burden.

6.2 Future Directions

Although millimeter-wave (mmWave) communications have emerged as one of the most promising technologies to support mobile data-intensive applications with different QoS requirements, there are still many open issues to be investigated. Next, we outline several important future research directions.

6.2.1 Beam Alignment Under High Mobility

It is widely acknowledged that the performance of mmWave communications significantly degrades in high mobility scenarios. The underlying reasons are twofold. Firstly, the beam direction changes with user mobility, which results in frequent beam misalignment and dis-connection between the transmitter and receiver. Secondly, mobile users also suffer from the blockage issue. For example, in vehicular mmWave networks, vehicular users not only need to frequently align their beams with roadside BSs but also deal with the LOS obstructions caused by either buildings or other vehicles (e.g., buses and trucks). Thus, the established mmWave link is intermittent and short-lived, which leads to a high probability of communication outage. It is critical to establish reliable mmWave communication links in mobile scenarios. As high-mobility users usually move along a specific trajectory, such as vehicles usually follow the road lane, such trajectory information can be exploited to enhance mmWave communications in high-mobility scenarios. For example, based on the velocity and moving direction of a vehicle, BS can predict the future locations of the vehicle so as to proactively align their beams before the link outage. However, as the real-time moving direction and location information are usually unknown to the BS *a priori*, an efficient and effective prediction scheme is necessary to optimally determine the beam direction based on the trajectory feature of mobile users and meanwhile maximize their quality of service (QoS).

6.2.2 Efficient QoS-Aware MAC Protocol

Currently, the physical layer techniques of mmWave networks have been well studied, e.g., BF. However, the study from the perspective of MAC layer is rather limited. Even with improved physical layer schemes, a coarse MAC protocol would still result in poor network performance. The design of enhanced MAC protocols becomes a new challenge for the entire mmWave network due to the following two reasons: Firstly, based on our analytical results in Chap. 4, the MAC throughput in the BF training stage degrades significantly in high-density scenarios, which limits the application of mmWave networks in dense-user scenarios. Secondly, since the type of data traffic becomes diverse, future mmWave networks should support heterogeneous traffic while satisfying various levels of QoS requirements. However, the MAC protocols in mmWave networks are not only inefficient, but also do not consider the diversified requirements for different services. Thus, designing an efficient and QoS-aware MAC protocol is of paramount importance for future mmWave networks.

6.2.3   Blockage-Aware mmWave Network

The blockage problem is regarded as one of thorny open issues in mmWave networks, especially in indoor scenarios. The mmWave connections can be easily suppressed when the LOS path is blocked by either the human body or indoor infrastructure (e.g., pillar and desk), which results in high link outage probability and unstable connections. When link outage occurs, the link has to be re-established, which results in frequent beam realignment at the cost of increased latency. In addition, the consequence of latency caused by the link outage is amplified at higher network layers. For example, at the transport layer, link outage may lead to timeout, and the transmission control protocol (TCP) connections are thus re-established, which further exacerbates the latency. Since a mobile user cannot always be served by one BS, exploring a user-centric framework that simultaneously serves a mobile user with multiple BSs is a practical solution. Specifically, when the system detects the connection between a mobile user and a BS is blocked, it proactively transfers to a new connection between the mobile user and another BS. In this way, the latency caused by link establishment can be ignored since the backup connection is established in time. However, how to accurately and proactively identify blockage is a challenging issue. In addition, developing an effective and low complexity link re-establishment solution between mobile users and BSs requires further investigation.

Printed in the United States
by Baker & Taylor Publisher Services